Horizontal Drilling Engineering - Theory, Methods and Applications

Horizontal Drilling Engineering - Theory, Methods and Applications

Editor

M. S. Chauhan

scitus
academics

Horizontal Drilling Engineering - Theory, Methods and Applications
Edited by **M. S. Chauhan**

Printed in 2017

ISBN: 978-1-68117-399-3

Library of Congress Control Number: 2015941589

© 2016 by

SCITUS Academics LLC,
616, Corporate Way, Suite 2, 4766,
Valley Cottage, NY 10989

www.scitusacademics.com

Contents

Preface

Drilling engineering is a challenging discipline in the oil patch. It goes beyond what is found in textbooks. The technological advances in the past two decades have been very significant. These advances have allowed the oil industry worldwide to economically and successfully exploit oil and gas fields that may have not been possible before. The fundamentals of fluid mechanics and solid mechanics, along with the basic scientific concepts of chemistry, form the basis of drilling engineering. The rewards and successes of drilling projects are predicated on the ability of the drilling engineer who fully understands all the engineering aspects and equipment required to drill a usable hole at the lowest dollar per foot, in vertical well drilling, or at the highest equivalent barrel of oil per foot in horizontal/multilateral well drilling. Horizontal Drilling Engineering book gives the fundamentals and field practices involved in horizontal drilling operations. Key Features & Benefits: This textbook is an excellent resource for drilling engineers, directional drillers, drilling supervisors and managers, and petroleum engineering students.

Editor

Chapter 1

Nanofabrication with Pulsed Lasers

AV Kabashin[1], Ph Delaporte[1], A Pereira[1, 2],
D Grojo[1], R Torres[1], Th Sarnet[1], and M Sentis[1]

[1]Lasers, Plasmas et Procédés Photoniques (LP3, UMR 6182 CNRS),
Université de la Méditerranée, Campus de Luminy-case 917,
13288, Marseille Cedex 9, France

[2]Laboratoire de Physico-Chimie des Matériaux Luminescents,
LPCML (UMR 5620 CNRS), Domaine Scientifique de la Doua,
Université Claude Bernard Lyon 1, 10 rue Ada Byron, 69622,
Villeurbanne, France

ABSTRACT

An overview of pulsed laser-assisted methods for nanofabrication,
which are currently developed in our Institute (LP3), is presented.

The methods compass a variety of possibilities for material nanostructuring offered by laser–matter interactions and imply either the nanostructuring of the laser-illuminated surface itself, as in cases of direct laser ablation or laser plasma-assisted treatment of semiconductors to form light-absorbing and light-emitting nano-architectures, as well as periodic nanoarrays, or laser-assisted production of nanoclusters and their controlled growth in gaseous or liquid medium to form nanostructured films or colloidal nanoparticles. Nanomaterials synthesized by laser-assisted methods have a variety of unique properties, not reproducible by any other route, and are of importance for photovoltaics, optoelectronics, biological sensing, imaging and therapeutics.

INTRODUCTION

When nanostructured, many materials start to exhibit new optical properties making them unique for a plethora of applications. In particular, despite small and indirect band gaps in the bulk state, the nanostructured IV group semiconductors (e.g. Si, Ge) become efficient size-dependent emitters in the visible light range [1,2], but also can work as photosensitizers to generate singlet oxygen under photoexcitation [3,4]. Another prominent example relates to noble metal nanostructures, which provide a number of unique plasmonic effects, including size-dependent absorption peaks [5,6], drastic local electric field enhancement [7,8], resolution beyond the diffraction limit [9], nanotrapping[10] etc. These new properties of emerging nanomaterials appear to be extremely promising for photovoltaics and optoelectronics, as well as for biological sensing, imaging and therapeutics.

The employment of pulsed lasers offers a novel unique tool for nanofabrication [11]. When focused on the surface of a solid target, pulsed laser radiation causes a variety of effects, including heating, melting and finally ablation of the target and such processes can lead to an efficient material nanostructuring, as shown in Fig. 1. First, the laser-assisted removal of material from the irradiated spot can result in a spontaneous formation of variety of periodic micro-

and nanoarchitectures within this spot [12-16]. Second, laser ablation of material from a solid target leads to the production of nanoclusters [17-20]. When produced in gaseous environment or in vacuum, these nanoclusters can then be deposited on a substrate yielding to the formation of a nanostructured film [17,21-24]. When produced in liquid environment, the nanoclusters can be released into the liquid forming a colloidal nanoparticle solution [25-32]. In all cases, properties of formed nanostructures can be unique and not reproducible by any other route [27-33]. As an example, the fabrication of nanoparticles in aqueous solutions does not require any chemical reducing agent, which conditions unique surface chemistry and purity of produced nanomaterials [28,29]. Furthermore, when synthesized in clean, biocompatible environment, laser-synthesized nanomaterials are exempt of any residual toxicity that is typical for chemically synthesized nanoparticles [32,33].

In this paper, we review laser-assisted technologies, developed by LP3 members, which are now available in our Institute.

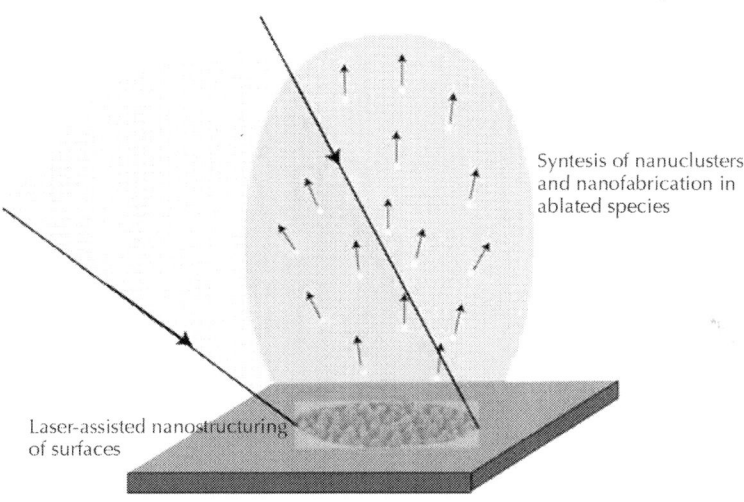

Syntesis of nanuclusters and nanofabrication in ablated species

Laser-assisted nanostructuring of surfaces

Figure 1: Schematics of laser–target interaction and material nanostructuring.

LASER-ASSISTED NANOSTRUC-TURING OF SURFACES

Properties of laser-materials interaction are known to be strongly dependent on parameters of laser radiation. Among these parameters, the wavelength and pulse length are especially important to determine the efficiency of radiation absorption, dynamics of plasma plume expansion and nanoclustering [11]. Ultraviolet, visible and near-infrared lasers are normally considered as most adequate for laser ablative tasks. Indeed, most industrially important materials efficiently absorb laser radiation in the spectral range of 200–1,000 nm, yielding to material ablation and nanostructuring. In addition, radiation of UV and visible lasers is relatively transparent to laser plasma, which minimizes laser beam distortions and power losses before reaching the target. In contrast, the interaction of infrared radiation (1–11 μm) with materials is characterized by a strong energy absorption by plasma itself, which completely changes conditions of nanostructuring. The pulse length is another important parameter in laser–matter interaction. Micro- and nanosecond laser–matter interactions are typically associated with long ablation regime, in which the ablation process takes place during the laser pulse itself. In contrast, pico- and femtosecond lasers offer short ablation regimes, in which moments of radiation absorption and material ablation are temporally separated. Although multi-pulse laser ablation from the target surface is always accompanied by the formation of micro- and nanoscale features and periodic structures [12-16,34], properties of these features strongly depend on the wavelength and pulse length. Therefore, depending on the task, one can select appropriate radiation parameters to condition the ablation regime and obtain prescribed properties of laser-synthesized nanostructures. Below, we give several examples of laser-assisted methodologies to fabricate nanostructures within the irradiation spot.

Femtosecond Laser Ablation of Si: Formation of Black Si for Photovoltaics Applications

It is accepted that femtosecond pulses give at least two major advantages to micromachining compared to nanosecond and longer pulses [35,36]: (1) the reduction of the pulse energy which is necessary to induce ablation for fixed laser wavelength and focussing conditions and (2) a significant reduction or complete removal of heat-affected zone (HAZ) and, as consequence, the improvement of the contour sharpness for the laser-processed structures. The second advantage is a direct consequence of the pulse being shorter than the heat diffusion time, given by the phonon transport. As in the case of nanosecond- and microsecond laser ablation, multi-pulse femtosecond ablation leads to a spontaneous formation of nanoarchitectures [14,37-40]. However, in contrast to long ablation, the fs regime is characterized by the absence of target melting effects, yielding to the formation of clean micro- and nanoscale features. In particular, using multi-pulse fs ablation of Si in the presence of SF_6 reactive gas, Mazur et al. [14,37] managed to fabricate extremely narrow micro-spikes within the irradiation spot, which are capable of efficiently absorbing light in the visible and infrared ranges. The efficient absorption of light exceeding 95% in the visible was attributed to a geometric multi-reflection effect offered by a unique spike-based structure, while the enhanced absorption in the infrared was explained by a sulphur doping [37]. Due to the wide-range absorption effect, the produced spike-based structure was called "black silicon" and was later used for the development of Si-based photodetectors. It is worth noting that such absorptive features cannot be reproduced by any alternative non-laser route. Other studies (see, e.g. [41]) reported the fabrication of nanostructured metal films exhibiting colours ("coloured metals") using similar fs ablation approach.

Our sub-project is devoted to the fabrication of "black silicon" structures for photovoltaics solar panel applications. The choice of photovoltaics as target application imposes new criteria on nanostructuring conditions. First, these applications require a high absorption mostly in the visible—near-IR range (300–1,000 nm), which enables us to exclude the necessity of using sulphur-based doping species. Second, these applications require uniform high-quality doping of nanostructured layers to maximize the photovoltage response. We succeeded in developing of a novel methodology to produce "black silicon" with such parameters, employing a Ti:Sapphire laser (wavelength 800 nm, pulse energy 5 mJ, repetition rate 1 kHz) [42]. In contrast to [14,37], we carry out multi-pulse laser processing in vacuum under the residual pressure of $(1-5) \times 10^{-5}$ mBar.

In addition, we avoid the doping procedure during the laser processing process and do it afterwards. To achieve a high quality doping of deep layers, the laser-structured samples are boron implanted by Plasma immersion (PULSION, BF3, 2 kV, 900°C, 30 min) and thermally annealed (TA). The junction depth obtained by this method is estimated to be about 150 nm, which is much shallower than the 3D laser structures; therefore, the junction follows the topography of the structures. Figure 2a demonstrates a silicon wafer surface after the laser processing and boron implantation procedure. Here, a rectangular area of 3×2 cm² is written by a programmed displacement with the speed of 150 µm/s of a femtosecond laser beam having the spot size of 35×35 µm². One can clearly see a black area on the silicon wafer associated with "black silicon".

As shown in Fig. 2b, the treated surface contains penguin-like nanospikes with the length of up to 10 µm and sub-µm lateral dimensions. Although the morphology of femtosecond laser-treated surface is rather different compared to narrow spike-like structures in [14,37], it is also characterized by an enhanced absorption in the visible range exceeding 90% (Fig. 2c). Depositing grating-like contacts on the top on the treated area, we were able to obtain the amplification of photocurrent by 50% compared to the untreated surface area. Such result was attributed to an enhanced absorption granted by the penguin-like structures, much larger surface of nanostructured silicon used for signal collection, and high quality of boron implantation offered by the post-ablation plasma implantation procedure. The fabricated structures are now actively tested as photovoltaics solar cells.

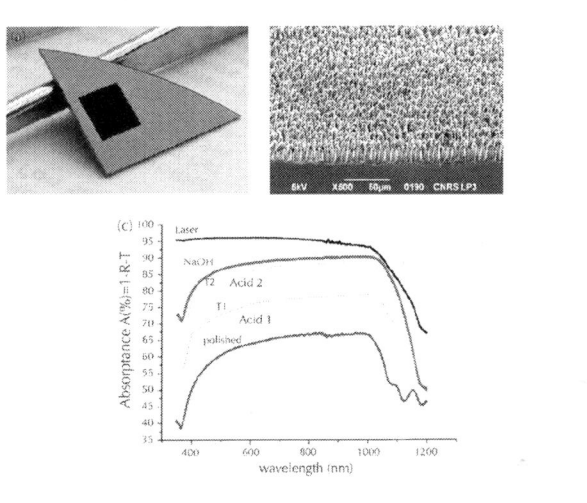

Figure 2: a Typical image of "black silicon" spot fabricated on a Si wafer by multi-pulse fs laser ablation in vacuum; b Typical scanning electron microscopy (*SEM*) image of penguin-like structure of black silicon; c Typical absorption spectra from "black silicon" and silicon treated by different methods.

Laser Plasma-assisted Nanostructuring of Surfaces

As we mentioned above, UV or ultrashort lasers contribute to a good radiation absorption by the target itself, while plasma remains relatively transparent to the incoming radiation. Such parameters ensure good quality of surface treatment in laser processing tasks. The plasma effect can be further minimized by reducing the pressure of the ambient gas. Depending on plasma plume size conditioned by the ambient gas pressure, the material can be re-deposited either within the irradiation spot (for high atmospheric pressure) or into the environment (for reduced pressures). In particular, for atmospheric pressure, the ablation process results in the formation of a deep crater, containing microscale spikes, covered by re-deposited nanoparticles [13-16]. In this case, chemical transformations in ablated species are minimal, since the ablated material rapidly cools down while interacting with the environment [24,43].

We recently introduced a novel method for surface nanostructuring, which is characterized by radically different nanofabrication conditions [44-49]. The method may look paradoxical, since it disaccords with main principles of laser processing requiring the minimization of plasma-related effects as one of main conditions to achieve high quality of laser treatment. In contrast, in this method, plasma-associated effects are amplified by all possible means. Basically, we use infrared radiation from CO_2 laser, which is strongly absorbed by the plasma itself. When focused in air and any other gas having atmospheric pressure, infrared radiation is capable of efficiently igniting the gas breakdown and this phenomenon is called the "laser spark". The presence of a target decreases the breakdown threshold by 2–3 orders of magnitude [50]. In the latter case, the target serves to provide first electrons. Then, an avalanche plasma discharge develops in ambient air moving towards the focusing lens. Absorbing main radiation power through the inverse Bremsstrahlung mechanism, the plasma accumulates an enormous amount of energy and is supposed to radically change conditions of nanocluster production and growth [51]. Indeed, in contrast to conventional laser ablation, the ablated species find themselves in a plasma "reactor" with extremely high temperatures (10^4 K) [52] and strong electromagnetic fields [53-56], yielding to a deep chemical transformation of properties of ablated clusters. The clusters then move back to the irradiated spot forming a film of clearly separated and densely packed spherical nanoparticles, as shown in Fig. 3a. The size of nanoparticles can vary for different materials, but is usually between 20 and 70 nm. The increase of plasma intensity can also lead to a coagulation of nanoparticles and the formation of much larger microscale spherical features. Nanostructures treated by this method have a specific texture with separated densely packed crystalline nanoparticle constituents, which contribute to unique optical properties. In particular, in the case of the laser plasma-based treatment of Zn in ambient air, the produced ZnO nanostructures exhibit very strong exciton-related peak around 380–385 nm under photoexcitation, whereas photoluminescence peaks associated with defects are essentially absent[49]. Furthermore, such nanostructure is capable of providing the mirror-less random lasing effect, arising as a result of a simultaneous strong amplification and scattering in a highly disordered medium [48]. Such effect is normally observed by the appearance of several extremely narrow lines within the exciton emission band under the increase of the pumping laser power. In the case of Si and Ge, the laser plasma treatment leads the formation of nanostructures, which are capable of generating strong photoluminescence (PL) in the visible [44-47]. Figure 3b shows PL properties of Si nanostructures fabricated by the laser plasma-based treatment. One can see two PL bands around 2.1 and 3.25 eV, associated with Si-based nanostructures. In the case of Ge, the PL bands are slightly different and situated around 2.2 and 2.9 eV [47].

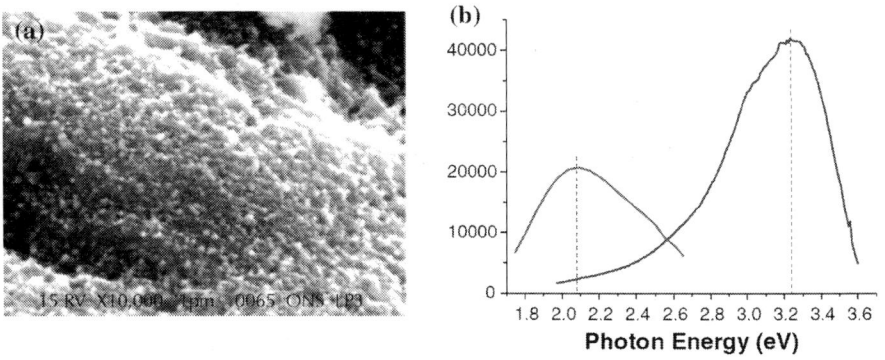

Figure 3: a Typical image of Si-based nanostructures prepared by laser plasma-assisted treatment of a Si wafer; b Photoluminescence spectra from laser plasma-treated nanostructured Si spots for different pumping wavelengths (325 and 488 nm).

Near-field Nanoparticle-assisted Nanostructuring of Surfaces: Fabrication of Patterned Nanoarrays

This sub-project addresses the formation of periodic nanoarrays by laser-assisted methodologies. The methodology implies two steps [57]: (1) laser ablation to form programmed periodic nanoscale features on a sacrificial surface layer; (2) post-ablation deposition/ chemical treatment step to fabricate nanoarrays. The first step is based on the use of near-field particle-assisted ablation to produce nanoscale features on various substrates [58-61]. A monolayer of self-assembled SiO_2 spheres is formed on a 20-nm alumina (Al_2O_3) film, as shown schematically in Fig. 4a. Then, pores are optically drilled in the Al_2O_3 film by particle-assisted near-field enhancement. This is accomplished through the illumination of the spheres with a single nanosecond laser pulse at the wavelength of 193 nm. Such process leads to the local removal of the 20-nm-thick Al_2O_3 film under each sphere. Since the spheres are arranged in a hexagonal array at the surface of the substrate, the Al_2O_3 film is decorated with an ordered arrangement of holes (Fig. 4b). The second step employs the laser-fabricated porous alumina membrane (LF-PAM) as a mask for the deposition of metal (Fig. 4c). Then, the alumina layer is dissolved yielding to the production of a series of ordered metal nanodots on the surface of the substrate (Fig. 4d). One of the main advantages of the proposed method consists in a possibility of producing nanodot arrays of functional materials, independently of the nature of the substrate.

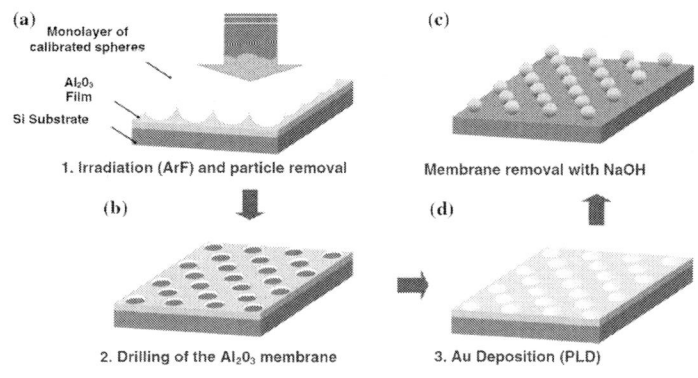

(a) Monolayer of calibrated spheres
Al₂O₃ Film
Si Substrate
1. Irradiation (ArF) and particle removal

(b)
2. Drilling of the Al₂O₃ membrane

(c)
Membrane removal with NaOH

(d)
3. Au Deposition (PLD)

Figure 4: Nanodot array fabrication. a A monolayer of spheres deposited on a thin alumina film is illuminated with a single laser pulse. b Near-field enhancement underneath the spheres leads to the parallel nanodrilling of the film. c A metal (gold in our case) is then deposited and the alumina membrane is dissolved in basic solution. d An ordered Au nanodot array is then obtained on the silicon substrate.

In particular, the proposed methodology can be used to fabricate an ordered array of gold nanodots (plasmonic arrays). Figure 5a shows an image of an Al_2O_3 thin film after drilling holes by 250-nm silica particle-assisted laser ablation. One can see that the laser-drilled holes are relatively uniform with the mean size of holes of about 100 nm, while the distance between the nanoholes is well conditioned by the size of self-assembled silica microparticles. As shown in Fig. 5b, 5c, the second deposition/chemical treatment step leads to the formation of high quality plasmonic arrays based on gold nanodots with the size of features less than 100 nm. These nanodot arrays are of importance for biosensing applications [62,63].

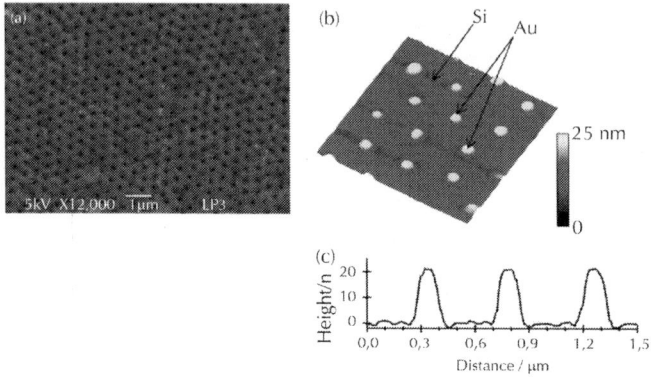

Figure 5: a SEM image of an Al_2O_3 thin film (20 nm) deposited on a Si substrate and simultaneously drilled by the near-field enhancement of a single nanosecond laser pulse, which is produced by a lattice of SiO_2spheres ($R = 250$ nm); b AFM image and **c** depth profile of gold nanodots created on silicon substrates by the LF-PAM-based process.

LASER ABLATIVE SYNTHESIS OF NANOCLUSTERS

The evolution of properties of formed nanostructures after the ablation process is mainly determined by the interaction of the species with the environment. In fact, in the first approximation, the nanocluster formation process can be described by the classical theory of condensation and nucleation in a vapour layer [64,65]. However, the growth of nanoclusters strongly depends on initial parameters of ablated species (energy, angular distribution, density) and by laser interaction with the environment, which are conditioned by characteristics of pumping radiation. In particular, the radiation can directly affect the nucleation process if the pumping laser pulse is long enough. Indeed, the energy of photons can be sufficient to produce nucleation centres, change the dynamics of the nuclei growth and modify the diffusion of species in the vapour phase. Below, we give a review of experimental results on the laser ablative nanostructure growth.

Laser Ablation in Residual Gases and Deposition of Nanostructured Films

The first possibility is related to the ablation of material in gaseous environment and the deposition of ablated species onto a substrate in pulsed laser deposition (PLD) geometry. In this geometry, the employment of a UV radiation from an excimer laser is normally preferable [11]. UV radiation is well absorbed by most industrially important materials, while the formed plasma plume is relatively transparent to it. The material can be in general ablated in vacuum; but due to a low probability of nanocluster coalescence in vacuum, it is difficult to control their growth and the nanoclusters normally deposit as ablated, forming a dense film with a significant amorphous fraction. Therefore, it is important to have a residual neutral light gas to affect the growth procedure. In this case, nanoclusters cool down under collisions with gas molecules or atoms, which contribute to their coalescence in the vapour phase. Under such conditions, the nanocluster growth process can be efficiently controlled by varying the pressure of the ambient gas [23,66,67].

To fabricate Si-based nanostructured films, we normally use radiation of a pulsed ArF or KrF lasers (193 or 248 nm, respectively, 15 ns FWHM, repetition rate 30 Hz) to ablate material from a rotating Si target. The radiation is focused at the incident angle of 45° to the surface. A substrate is placed on a rotating holder in front of the target. The experimental chamber is filled with helium for a deposition at a constant pressure ranging between 0.05 and 10 Torr. The film thickness after several thousands laser shots is 100–700 nm.

Figure 6 shows a transmission electron microscopy (TEM) image of several isolated nanoparticles, synthesized by laser ablation from a Si target and deposited on a carbon-coated TEM grid (a) and corresponding nanoparticle size distribution (b). One can see that the produced nanoclusters are very small with the size in the range of 1–4 nm. As shown in Fig. 6c, the mean size of nanoparticles slightly increases under the increase of He pressure. For example, the increase of the pressure from 0.5 to 8 Torr results in the increase of the nanoparticle size from 1.5 to 4 nm. Another important feature relates to essentially porous texture of the films prepared by pulsed laser deposition, as illustrated by the inset of Fig. 6c, and the porosity of films increases with the increase of the ambient gas pressure (Fig. 6c). Indeed, while the deposition under 1 Torr results only in some germs of roughness, the experiment under 2 Torr provides a developed porous structure with pore size of about 50–100 nm. A further pressure increase up to 4 Torr leads to a formation of web-like aggregations of particles. As shown in Fig. 6c, the deposition of films at high pressures (>4 Torr) leads to porosities exceeding 90%, corresponding to the formation of powders on the substrate. Thus, the pressure of the ambient gas appears to be one of key parameters, which determines both the size of synthesized nanocrystals and the porosity of deposited nanostructured layers. Such a strong impact of the gas pressure on nanoclustering process suggests the importance of cooling of ablated species under their collisions with gas atoms. If the pressure of the gas is high enough, the nanoclusters experience a sufficient number of collisions to rapidly cool down and crystallize in the vapour phase. As a result, they arrive on the substrate in the form of a powder. In contrast, low collision regime at low pressures advantages the formation of dense and low-porous films.

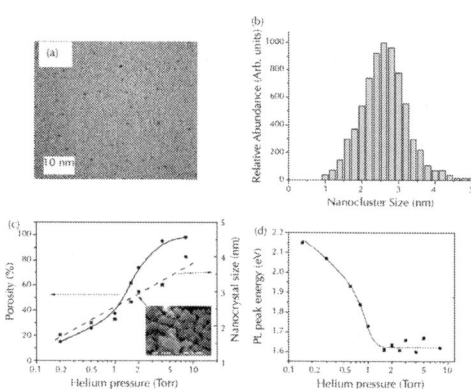

Figure 6: Transmission electron microscopy image of Si nanoparticles fabricated by pulsed laser ablation at 2 Torr of He (a) and corresponding

nanocluster size distribution (b); c Dependence of the nanocrystal size (dashed line) and film porosity (solid line) on the pressure of He during the deposition; Inset typical scanning electron microscopy image of films prepared by pulsed laser ablation; d Dependence of the position of PL peak from laser-ablated films on the gas pressure during the deposition.

It is important that all Si-based films prepared by the PLD method exhibit strong visible emission (PL) under photoexcitation, while PL properties of the films are quite different for films of different porosities. Low-porous films $P < 40\%$, deposited at reduced pressures $P < 1.5$ Torr, exhibit relatively weak PL with peak energy strongly depending on the gas pressure. In particular, the pressure decrease from 1.5 to 0.15 Torr in different depositions causes a blue shift of the peak from 1.6 to 2.15 eV, as shown in Fig. 6d. In contrast, films with an enhanced porosity $P > 40\%$, deposited at higher pressures, provide only spectra with fixed peaks. The first band (1.6–1.7 eV) is independent of the pressure and can be seen just after the fabrication process. An additional 2.2–2.3 eV band can appear under the oxidation of samples in humid air. Here, PL properties of low-porous films (porosity <40%) are almost unchangeable under these conditions, while the integral PL intensity from highly porous films significantly increases with the prolonged oxidation. We attribute such a difference of PL properties to the impact of post-fabrication natural oxidation, controlled by the level of porosity. Dense and self-coagulated structures of the films fabricated under $P < 1$ Torr minimize the impact of ambient atmosphere on the film properties; and for these films, mechanisms related to core silicon crystals became predominant. Since the blue shift of the spectra under the decrease of helium residual pressure is accompanied by a certain decrease of the nanocrystal size, the quantum confinement mechanism [1] can be considered as one of main opportunities. In contrast, a high porosity enhances the surface area, which is subjected to surface chemistry modifications due to interactions of nanocrystallites with oxygen. This can drastically enhance the role of oxidation in the formation of PL centres and the domination of oxygen-related PL mechanisms connected either to defects in the SiO_2 structure (usually, this mechanism provides 2–2.4 eV PL [68]) or to the interfacial layer (1.65 eV) [69].

Ultra-short Laser Ablation in Liquid Environment to Form Colloidal Nanoparticles

In the case of liquid ambience, laser ablation process leads to the release of nanoclusters into the liquid and the formation of a colloidal nanoparticle solution [11]. In contrast to conventional chemical reduction methods, this method enables avoidance of the use of toxic chemical reduction agents to control the growth of the nano-particles. As an example, the laser ablation-based synthesis can be implemented in pure deionized water. The independence of laser-based synthesis of dirty colloidal chemistry makes it unique for the fabrication of markers of bioanalytes for sensing and imaging applications. In pure water, however, or any other aqueous solution exempted of additional chemically active components, the size of nanoparticles produced tends to be relatively large, since a natural coalescence of hot ablated nanoclusters cannot be easily overcome. In particular, nanosecond laser-based ablation used in most works, generally gives relatively large (10–300 nm) and strongly dispersed (50–300 nm) particles [25-28]. Although certain size control can be achieved by decreasing the wavelength of pumping radiation or varying the laser fluence, the range of size variations stays rather moderate in the nanosecond pulse case. Mafune [27] showed that size characteristics of nanoparticles can be improved by adding some reactive surfactants such as sodium dodecyl sulphate (SDS) or CTAB during the ablation. As an example, thiol-containing SDS covered gold nanoclusters just after their production and thus protected them from further coalescence. However, bioimaging applications of so produced nanoparticles are hardly possible since the surfactants are not biocompatible.

We recently proposed a femtosecond laser ablation-based method for nanoparticle synthesis, which makes possible an efficient control of size of prepared nanoparticles by varying physical parameters[28-33]. The experiments are normally carried out with an Ytterbium (400 fs FWHM, 1,025 nm, 1 kHz) or a Ti/

Sapphire laser (110 fs FWHM, 800 nm, 1 µJ/pulse, 1 kHz). The radiation is focused onto a target of different materials (Au, Ag, Si, Ti, Cu), which is placed on the bottom of a glass vessel filled with aqueous solutions. The vessel is placed on a horizontal moving platform to avoid the ablation of material from the same area. The ablation experiments are carried out in pure deionized water and in aqueous solutions (biopolymers, cyclodextrins).

When performed in relatively neutral environment such as deionized water, fs laser ablation normally leads to the formation of two nanoparticle populations, independently on the material of the target, as shown in Fig. 7. The first population is characterized by a small mean nanoparticle size and narrow size dispersion, whereas the second one has a much larger mean size and broader size dispersion. The presence of the two populations suggests the involvement of two different mechanisms of nanoparticle growth. The production of the first, less dispersed population is characterized by the absence of target melting effects, suggesting direct radiation-related ablation of material [28]. In contrast, the production of the highly dispersed population is accompanied by a strong melting of material inside the ablated crater. This melting is usually attributed to the explosion of a cavitation bubble formed as a result of energy transfer from laser plasma to the liquid [11]. It is important that in the fs laser ablation regime, the nanoparticle size can be efficiently controlled by varying the intensity of radiation during the nanosynthesis process. In particular, the mean size of gold nanoparticles can be reduced from 120 to 4 nm by the decrease of laser fluence down to the threshold values [28]. Similar effect can be achieved by changing the radiation focusing on the target surface [29]. Such efficient method of size control, not possible with nanosecond or microsecond pulses, is granted by specific conditions of fs laser–materials interaction. In the fs ablation regime, much less radiation energy is transferred to the cavitation bubble (15% compared to 80% in the nanosecond pulse regime [70]). We believe that the decrease of the laser fluence down to near-threshold values enables one to completely avoid cavitation phenomena and thus eliminate the second highly size-dispersed nanoparticle population. In addition, the decrease of the laser fluence changes parameters of ablated nanoclusters (energy, angular distribution etc.), which can in turn affect the final size of nanoparticles of the first population. In particular, using this method in the case of gold, we managed to vary the size of synthesized nanoparticles between 4 and 20 nm [28].

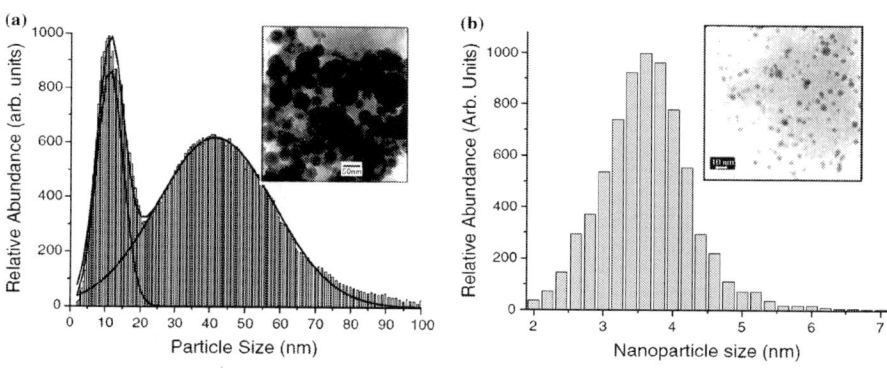

Figure 7: Transmission electron microscopy images and corresponding size distributions of a TiOx nanoparticles prepared by fs laser ablation from a Ti target in deionized water; b gold nanoparticles prepared in aqueous solution of polyethylene glycol.

Another important issue is related to chemical properties of laser-synthesized nanoparticles. Since the nanoparticles are produced by pure physical ablation from a target without the involvement of any specific chemicals, surface chemistry of these nanoparticles can drastically differ from that of counterparts prepared by conventional colloidal chemistry [30,31,71-73]. In particular, laser-synthesized gold becomes susceptible to oxidation and, in contrast to chemically prepared gold, the surface of these nanoparticles is partially covered by a layer of oxide. Furthermore, the oxidized nanoparticle surface can have different termination, depending on the pH of the environment. The oxidized portion of the gold surface normally has Au–O– groups at pH > 5.8 and increasing numbers of Au–OH groups at pH < 5.8 [71]. This oxide-related surface termination makes possible interactions of gold nanoparticles with groups, for which conventional gold is normally inert. A prominent example of new gold chemistry is related to the use of biopolymers [72] and oligosaccharides [30]. Although these substances do not contain gold-reactive thiol group, they react with laser-synthesized gold nanoparticles, yielding to a drastic reduction of the nanoparticle size. Figure 7b illustrates the effect of the reduction of nanoparticle mean size under the use of a biopolymer polyethylene glycol (PEG). One can see that the nanoparticles' mean size and size dispersion can be reduced down to 3 nm with the size dispersion not exceeding 1.5 nm FWHM. We believe that such reduction of nanoparticle size is a result of the hydrogen bonding of the –OH groups of these compounds and the $–O^-$ at the gold surface. The molecules of PEG cover gold nanoclusters just after ablation and act like "bumpers", limiting contact between particles, preventing their coalescence (when the particles are still "hot") and aggregation (when the particles are "cold"). Similar mechanism takes place during laser ablation in aqueous solutions of other polymers (dextran etc.) [72] and oligosaccharides (cyclodextrin) [30]. Thus, OH groups of different biocompatible compounds can efficiently react with oxidized gold surface leading to the reduction of the nanoparticle size. It is important that in contrast to SDS or other surfactants previously used to control the nanoparticle size [27], biopolymers and oligosaccharides are essentially biocompatible. Moreover, the ultra-pure laser-ablated nanoparticles can be functionalized by a proper chemical modification of chemicals. We believe that this gives a huge advantage over the chemically produced nanoparticles for nano-engineering and functionalization of nanoparticles produced, as well as for a solution of toxicity problems. In particular, PEG is known as one of best materials to avoid the immune response in "in vivo" applications of inorganic nanoparticles. When covered by PEG, the nanoparticles become invisible for the immune system. In the case of laser-synthesized nanoparticles, one does not need to use any intermediate chemical group to link nanoparticles to PEG, as it takes place in the case of chemically synthesized nanoparticles.

CONCLUSIONS

We reviewed on the development, in our Institute, of various laser-assisted methodologies for nanofabrication in the gaseous and liquid environment. The methodologies imply the fabrication of nanoparticles/nanostructures either within the laser irradiation spot on the target surface or in the ablated species. Laser-synthesized nanomaterials exhibit unique optical properties and are exempt of toxicity, which make them very important for photovoltaics, optoelectronics, biological sensing, imaging and therapeutics.

ACKNOWLEDGMENTS

The authors are grateful to Agence Nationale de Recherche (ANR) and Ion Beam Services (IBS company) for Plasma Immersion doping of the black silicon.

REFERENCES

1.   Canham LT:*Appl. Phys. Lett.*. 1990, 57:1046.COI number [1:CAS:528:DyaK3cXmt1ahurc%3D];   Bibcode   number [1990ApPhL..57.1046C]

2.   Cullis AG, Canham LT, Calcott PD:*J. Appl. Phys.*. 1997, 82:909.COI number [1:CAS:528:DyaK2sXkvFCntbg%3D]; Bibcode number [1997JAP....82..909C]

3.   Kovalev D, Gross E, Künzner N, Koch F, Timoshenko VY, Fujii M:*Phys. Rev. Lett.*. 2002, 8913:137401.Bibcode number [2002PhRvL..89m7401K]

4.   Timoshenko VY, Kudryavtsev AA, Osminkina LA, Vorontsov AS, Ryabchikov YV, Belogorkhov IA, Kovalev D, Kashkarov PK:*JETP Lett.*. 2006, 83:423.COI number [1:CAS:528:DC%2 BD28XmslSnurs%3D]

5.   Kerker M: *The Scattering of Light and Other Electromagnetic Radiation*. Academic Press, New York; 1969.

6. Kreibig U, Vollmer M: *Optical Properties of Metal Clusters.* Springer, Berlin; 1996.

7. Nie S, Emory SR:*Science.* 1997, 275:1102-1106.COI number [1:CAS:528:DyaK2sXhtlGlsL4%3D]

8. Li K, Li X, Stockman M, Bergman D:*Phys. Rev. B.* 2005, 71:115409.Bibcode number [2005PhRvB..71k5409L]

9. Yao J, Liu Z, Liu Y, Wang Y, Sun C, Bartal G, Stacy A, Zhang X:*Science.* 2008, 321:930.COI number [1:CAS:528:DC%2B D1cXpslWrtLo%3D]; Bibcode number [2008Sci...321..930Y]

10. Grigorenko AN, Roberts NW, Dickinson MR, Zhang Y:*Nat. Photon..* 2008, 2:365-368.COI number [1:CAS:528:DC%2BD 1cXnsFKnsbo%3D]; Bibcode number [2008NaPho...2..365G]

11. Kabashin AV, Meunier M: Recent advances. In *Laser Processing Material.* Edited by Perriere J, Millon E, Fogarassi E. Elsevier, Amsterdam; 2006:1-36.

12. Prokhorov AM, Konov VI, Ursu I, Mikhailescu IN: *Laser Heating of Metals.* Hilger, Bristol; 1990.

13. Krajnovich DJ, Vazquez JE:*J. Appl. Phys..* 1993, 73:3001. COI number [1:CAS:528:DyaK3sXhvFalsrY%3D]; Bibcode number [1993JAP....73.3001K]

14. Her T-H, Finlay RJ, Wu C, Deliwala S, Mazur E:*Appl. Phys. Lett..* 1998, 73:1673.COI number [1: CAS: 528:DyaK1cXmtValtr4%3D]; Bibcode number [1998ApPhL..73.1673H]

15. Pedraza AJ, Fowlkes JD, Lowndes DH:*Appl. Phys. Lett..* 1999, 74:2322.COI number [1: CAS: 528:DyaK1MXit1yiurk%3D]; Bibcode number [1999ApPhL..74.2322P]

16. Costache F, Henyk M, Reif J:*Appl. Surf. Sci..* 2002, 186:352. COI number [1:CAS:528:DC%2BD38XitFart78%3D]; Bibcode number [2002ApSS..186..352C]

17. Movtchan IA, Marine W, Dreyfus RW, Le HC, Sentis M, Autric M:*Appl. Surf. Sci..* 1996, 96–98:251.

18. Geohegan DB, Puretzky AA, Duscher G, Pennycook SJ:*Appl. Phys. Lett..* 1998, 73:438.COI number [1:

CAS: 528:DyaK1cXksleisLg%3D];   Bibcode   number [1998ApPhL..73..438G]

19. Geohegan DB, Puretzky AA, Duscher G, Pennycook SJ:*Appl. Phys. Lett.*. 1998, 72:2987.COI number [1: CAS: 528:DyaK1cXjsVOhurY%3D];   Bibcode   number [1998ApPhL..72.2987G]

20. Povarnitsyn ME, Itina TE, Sentis M, Khishenko KV, Levashov PR:*Phys. Rev. B Phys. B*. 2007, 75:235414.Bibcode number [2007PhRvB..75w5414P]

21. Werwa E, Seraphin AA, Chiu LA, Zhou C, Kolenbrander KD:*Appl. Phys. Lett.*. 1994, 64:1821.COI number [1: CAS: 528:DyaK2cXjtFajtbc%3D];   Bibcode   number [1994ApPhL..64.1821W]

22. Yamada Y, Orii T, Umezu I, Takeyama Sh, Yoshida T:*Jpn. J. Appl. Phys.*. 1996, 35:1361.COI number [1:CAS:528:DyaK28XisFSqu70%3D];   Bibcode   number [1996JaJAP..35.1361Y]

23. Kabashin AV, Sylvestre J-P, Patskovsky S, Meunier M:*J. Appl. Phys.*. 2002, 91:3248.COI number [1:CAS:528:DC%2BD38 Xhs1agsr8%3D]; Bibcode number [2002JAP....91.3248K]

24. Pereira A, Cros A, Delaporte Ph, Georgiou S, Manousaki A, Marine W, Sentis M:*Appl. Phys. A*. 2004, 79:1433.COI number [1:CAS:528:DC%2BD2cXlslCiurw%3D]; Bibcode number [2004ApPhA..79.1433P]

25. Henglein AJ:*Phys. Chem.*. 1993, 97:5457.COI number [1:CAS:528:DyaK3sXisVChurc%3D]

26. Nedderson J, Chumanov G, Cotton TM:*Appl. Spectrosc.*. 1993, 47:1959.Bibcode number [1993ApSpe..47.1959N]

27. Mafuné F, Kohno J-Y, Takeda Y, Kondow T, Sawabe H:*J. Phys. Chem. B*. 2000, 104:8333.

28. Kabashin AV, Meunier M:*J. Appl. Phys.*. 2003, 94:7941.COI number [1:CAS:528:DC%2BD3sXps1Wltbk%3D]; Bibcode number [2003JAP....94.7941K]

29. Sylvestre J-P, Kabashin AV, Sacher E, Meunier M:*Appl. Phys.*

A. 2004, 80:753.Bibcode number [2005ApPhA..80..753S]

30. Kabashin AV, Meunier M, Kingston C, Luong JHT:*J. Phys. Chem. B.* 2003, 107:4527.COI number [1:CAS:528:DC%2B D3sXjtlWhsrw%3D]

31. Sylvestre J-P, Kabashin AV, Sacher E, Meunier M, Luong JHT:*J. Am. Chem. Soc. (Commun.).* 2004, 126:7176.COI number [1:CAS:528:DC%2BD2cXktVSlsL4%3D]

32. Kabashin AV, Meunier M:*J. Photochem. Photobiol. A.* 2006, 182:330-334.COI number [1:CAS:528:DC%2BD28Xot1Glsb 8%3D]

33. Besner S, Kabashin AV, Winnik FW, Meunier M:*Appl. Phys. A.* 2008, 93:955-959.COI number [1:CAS:528:DC%2BD1cXht1Knsb7P]; Bibcode number [2008ApPhA..93..955B]

34. Tokarev V, Marine W, Sentis M, Prat C:*J. Appl. Phys..* 1995, 77:4714.COI number [1:CAS:528:DyaK2MXlt1ClsLg%3D]; Bibcode number [1995JAP....77.4714T]

35. Chichkov BN, Momma C, Nolte S, Von Alvensleben F, Tünnermann A:*Appl. Phys. A.* 1996, 63:109.Bibcode number [1996ApPhA..63..109C]

36. Parisse JD, Marine M, Sentis J:*Phys. IV 9.* 1999, 9:PR5-149.

37. Wu C, Crouch CH, Zhao L, Mazur E:*Appl. Phys. Lett..* 2002, 81:1999.COI number [1: CAS: 528:DC%2BD38XmvVSht78%3D]; Bibcode number [2002ApPhL..81.1999W]

38. Reif J, Costache F, Henyk M, Pandelov SV:*Appl. Surf. Sci..* 2002, 197–198:891.

39. Daminelli G, Krüger J, Kautek W:*Thin Solid Films.* 2004, 467:334.COI number [1:CAS:528:DC%2BD2cXnsVGlsbs% 3D]; Bibcode number [2004TSF...467..334D]

40. Besner S, Degorce J-Y, Kabashin AV, Meunier M:*Appl. Surf. Sci..* 2005, 247:163-168.COI number [1:CAS:528:DC%2BD2 MXltVOgs7k%3D]; Bibcode number [2005ApSS..247..163B]

41. Vorobyev AY, Guo C:*Appl. Phys. Lett..* 2008, 92:041914.

Bibcode number [2008ApPhL..92d1914V]

42. Halbwax M, Sarnet T, Delaporte Ph, Sentis M, Etienne H, Torregrosa F, Vervisch V, Perichaud I, Martinuzzi S:*Thin Solid Films*. 2008, 516:6791.COI number [1:CAS:528:DC%2BD1 cXnvVKitb0%3D]; Bibcode number [2008TSF...516.6791H]

43. Pereira A, Delaporte Ph, Sentis M, Cros A, Marine W, Basillais A, Thomann AL, Leborgne C, Semmar N, Andreazza P, Sauvage T:*Thin Solid Films*. 2004, 453–454:16.

44. Kabashin AV, Meunier M:*Appl. Phys. Lett.*. 2003, 82:1619. COI   number   [1:CAS:528:DC%2BD3sXhvV2ltL4%3D]; Bibcode number [2003ApPhL..82.1619K]

45. Yang D, Kabashin A, Pilon-Marien V-G, Sacher E, Meunier M:*J. Appl. Phys.*. 2004, 95:5722.COI number [1:CAS:528:DC%2B D2cXjvVSrtL0%3D]; Bibcode number [2004JAP....95.5722Y]

46. Kabashin AV, Meunier M:*Mat Sci. Eng. B*. 2003, 101:60-64.

47. Kabashin AV, Magny F, Meunier M:*J. Appl. Phys.*. 2007, 101:054311.Bibcode number [2007JAP...101e4311K]

48. Kabashin AV, Trudeau A, Marine W, Meunier M:*Appl. Phys. Lett.*. 2007, 91:201101.Bibcode number [2007ApPhL..91t1101K]

49. Kabashin AV, Trudeau A, Marine W, Meunier M:*Appl. Phys. A*. 2008, 91:621.COI number [1:CAS:528:DC%2BD1cXlslKktb g%3D]; Bibcode number [2008ApPhA..91..621K]

50. Bunkin FV, Konov VI, Prokhorov AM, Fedorov VB:*JTP Lett.*. 1969, 9:371.Bibcode number [1969JETPL...9..371B]

51. Itina T, Hermann J, Delaporte Ph, Sentis M:*Appl. Surf. Sci.*. 2003, 128:27.

52. Raizer YP: *Laser-induced Discharge Phenomena*. Consultants Bureau, New York; 1977.

53. Drouet MG, Pepin H:*Appl. Phys. Lett.*. 1976, 28:426.COI number [1: CAS: 528:DyaE28XhslWqs7g%3D]; Bibcode number [1976ApPhL..28..426D]

54. Korobkin VV, Serov RV:*Pisma Zh. Eksp. Teor. Fiz.*. 1966, 4:103. [see also J. Exp. Theor. Phys. Lett. 4, 70 (1966)]

55. Kabashin AV, Nikitin PI:*Quantum Electron.*. 1997, 27:536.

Bibcode number [1997QuEle..27..536K]

56. Kabashin AV, Nikitin PI, Marine W, Sentis M:*Appl. Phys. Lett.*. 1998, 73:25.COI number [1: CAS: 528:DyaK1cXktlOhsbk%3D]; Bibcode number [1998ApPhL..73...25K]

57. Pereira A, Grojo D, Chaker M, Delaporte Ph, Guay D, Sentis M:*Small.* 2008, 4:572.COI number [1:CAS:528:DC%2BD1c XmvFWltLk%3D]

58. Piglmayer K, Denk R, Bauerle D:*Appl.Phys.Lett.*.2002,80:4693-4695.COI number [1: CAS: 528: DC%2BD38XksleltL8%3D]; Bibcode number [2002ApPhL..80.4693P]

59. Huang SM, Hong MH, Luk'yanchuk BS, Zheng YW, Song WD, Lu YF, Chong TC:*J. Appl. Phys.*. 2002, 92:2495-2500. COI number [1:CAS:528:DC%2BD38Xmt1Giu74%3D]; Bibcode number [2002JAP....92.2495H]

60. Huang SM, Sun Z, Luk'yanchuk BS, Hong MH, Shi LP:*Appl. Phys. Lett.*. 2005, 86:161911.Bibcode number [2005ApPhL..86p1911H]

61. Lu Y, Chen SC:*Nanotechnology.* 2003, 14:505-508.COI number [1: CAS: 528: DC%2BD3sXlslart7c%3D]; Bibcode number [2003Nanot..14..505L]

62. Anker JN, Hall WP, Lyandres O, Shah NC, Zhao J, Van Duyne RP:*Nature Mater.*. 2008, 7:442-453.COI number [1:CAS:528:DC%2BD1cXmsVejt7g%3D]; Bibcode number [2008NatMa...7..442A]

63. Kabashin AV, Evans P, Patskovsky S, Wurtz G, Hendren W, Dickson W, Pollard RJ, Podolsky V, Zayats AV:*Nature Mater.*. 2009, 8:867-871.COI number [1:CAS:528:DC%2BD1MXhtlSgs77O]; Bibcode number [2009NatMa...8..867K]

64. Abraham FF: *Homogeneous Nucleation Theory: The Pretransition Theory of Vapor Condensation.* Academic Press, New York; 1974.

65. Kashchiev D: *Nucleation: Basic Theory with Applications.* Butterworth-Heinemann, Oxford; 2000.

66. Kabashin AV, Charbonneau-Lefort M, Meunier M, Leonelli R:*Appl. Surf. Sci.*. 2000, 168:328.COI number [1: CAS: 528: DC%2BD3cXoslCqtbY%3D]; Bibcode number [2000ApSS..168..328K]

67. Kabashin AV, Meunier M, Leonelli R:*J. Vacuum Sci. Tech. B.* 2001, 19:2217.COI number [1:CAS:528:DC%2BD3MXptFWntLg%3D]

68. Prokes SM:*Appl. Phys. Lett.*. 1993, 62:3244.COI number [1: CAS: 528:DyaK2cXmsV2qsg%3D%3D]; Bibcode number [1993ApPhL..62.3244P]

69. Kanemitsu Y, Ogawa T, Shiraishi K, Takeda K:*Phys. Rev. B.* 1993, 48:4883.COI number [1:CAS:528:DyaK3sXmtlOjsLg%3D]; Bibcode number [1993PhRvB..48.4883K]

70. Vogel A, Noack J, Nahen K, Theisen D, Busch S, Parlitz U, Hammer DX, Noojin GD, Rockwell BA, Birngruber R:*Appl. Phys. B.* 1999, 68:271.COI number [1:CAS:528:DyaK1MXnsFOitg%3D%3D]; Bibcode number [1999ApPhB..68..271V]

71. Sylvestre J-P, Poulin S, Kabashin AV, Sacher E, Meunier M, Luong JHT:*J Phys Chem B.* 2004, 108:16864-16869.COI number [1:CAS:528:DC%2BD2cXotFSqsb0%3D]

72. Besner S, Kabashin AV, Meunier M, Winnik FM:*J Phys. Chem C.*. 2009, 113:9526-9531.COI number [1:CAS:528:DC%2BD1MXls1eitLw%3D]

73. Rioux D, Laferriere M, Douplik A, Shah D, Lilge L, Kabashin AV, Meunier M:*J. Biomed. Optics.* 2009, 14:021010.Bibcode number [2009JBO....14b1010R]

Chapter **2**

Review of Underground Coal Gasification Technologies and Carbon Capture

Stuart J Self, Bale V Reddy, and Marc A Rosen

Faculty of Engineering and Applied Science, University of Ontario Institute of Technology, 2000 Simcoe Street North, Oshawa, ON, L1H 7K4, Canada

ABSTRACT

It is thought that the world coal reserve is close to 150 years, which only includes recoverable reserves using conventional techniques. Mining is the typical method of extracting coal, but it has been estimated that only 15% to 20% of the total coal resources can be recovered in this manner. If unrecoverable coal is considered in the reserves, the lifetime of this resource would be greatly extended, by perhaps a couple hundred years. Mining involves a large amount of time, resources, and personnel and contains many challenges such as drastic changes in landscapes, high machinery costs, elevated risk to personnel, and post-extraction transport. A new type of coal extraction method, known as underground coal gasification (UCG), that addresses most of the problems of coal mining is being investigated and implemented globally. UCG is a gasification process applied to in situ coal seams. UCG is very similar to aboveground gasification where syngas is produced through the same chemical reactions that occur in surface gasifiers. UCG has a large potential for providing a clean energy source through carbon capture and storage techniques and offers a unique option for CO_2 storage. This paper reviews key concepts and technologies of underground coal gasification, providing insights into this developing coal conversion method.

REVIEW

Introduction

The global energy supply is comprised of many different sources, including fossil fuels, uranium, and various alternative and renewable sources. Currently, over 85% of the global energy supply is derived from fossil fuels, and a high fossil fuel dependency appears likely to remain in the immediate future [1]. The amount of energy required globally is projected to increase due to growing population and industrialization [2, 3]. Some feel that the total primary energy demand will double or even triple by the year 2050 relative to levels today, and as the energy demand continues to increase, future fossil fuel shortages are predicted [4,5].

Hammond [6] argues that fossil fuel depletion is a significant factor when considering sustainable energy systems for the future. Fossil fuel resources are finite and being consumed rapidly, beginning with the most economically attractive resources [7]. In the future, fossil fuel resource extraction and production rates are expected to peak and begin to decline [7]. Oil production is predicted to peak in 5 to 15 years and gas production within 40 years, with significant exhaustion of oil and gas reserves by the years 2050 and 2070, respectively [4,8]. As fossil fuel demand approaches supply levels, the cost of energy is anticipated to increase drastically, prompting research and technological developments for improved ways to convert more fossil fuel resources into useable reserves [9].

Currently, coal generates 41.5% of the world's electricity and provides 26.5% of global primary energy needs [1]. Coal has the largest reserves in the world of the fossil fuels and is abundant in many countries. It is thought that the world's recoverable coal reserve is close to 150 years at current production rates, but this only represents 15% to 20% of the entire resource [8]. Remaining global coal resources have recently been estimated to be 18 trillion tonnes [10]. This contrasts significantly with the typical figure of tens of billions of tonnes for recoverable reserves [4]. If unrecoverable coal is considered in the recoverable reserves, the lifetime of the resource could be extended by a couple hundred years. For this to be realized, new, economic extraction techniques need to be implemented.

Coal is conventionally extracted by mining, both underground and open pit. Mining operations require much time, personnel, and natural resources; typically, coal reserves lie too deep underground, or are otherwise too costly, to exploit using conventional mining methods. Conventional mining also has other issues including land subsidence, high machinery costs, hazardous work environments, coal transport requirements, localized flooding, and methane buildup in cellars of nearby homes [11].

Underground coal gasification (UCG) is a newer type of coal extraction that is being investigated and implemented around the world and that avoids most of the problems of mining coal. UCG involves the conversion of unmined coal, where coal seams are gasified, without mining, and synthetic gas (syngas) is produced for use in power generation or as chemical feedstock [12]. UCG limits the amount of underground work required by personnel, lowering risks of harm relative to conventional mining. Power generation and chemical processing plants can be built directly above a coal

Underground Coal Gasification

Brief UCG History

The concept of coal gasification has been recognized for more than 200 years and was first used during the late 1800's to produce town gas fuel for heating and lighting applications [14]. Today, coal gasification is primarily used to provide fuel for advanced power resource and use syngas produced through UCG, avoiding coal transport. UCG has the ability to significantly widen the resource base, where the energy contained within inaccessible coal reserves, considered uneconomical for recovery, could be recovered using UCG [11]. It has been estimated by the Underground Coal Gasification Partnership that around 4 trillion tonnes of otherwise unusable coal could be suitable for UCG [9].

UCG is appealing for expanding recoverable coal reserves, but as with the combustion of all fossil fuels, there are associated greenhouse gas emissions. Coal is the most carbon-intensive of all fossil fuels and has high associated CO_2 emissions [4]. The calorific value of fossil fuel sources varies, with typical values of 50 GJ/tonne for natural gas, 45 GJ/tonne for crude oil, and 30 GJ/tonne for coal [4]. Hence, coal has the highest CO_2 emissions per unit of thermal energy produced [13]. If coal is to become a major contributor in the future global energy supply, CO_2 capture and storage techniques would need to be incorporated in the process. UCG has good potential for CO_2 reduction. During gasification, CO_2 is produced, which can be captured from the syngas and stored for long terms. If UCG is successfully linked to such carbon capture and storage (CCS), a method will be available for exploiting the energy in previously unrecoverable coal reserves while satisfying standards for reducing CO_2 emissions.

The aim of this paper is to review key areas and technologies for underground coal gasification so as to provide insights into this developing coal conversion method.

plants and chemical feedstocks for use in the chemical industry [9]. Conventionally, coal is extracted from the ground through mining, processed, transported, and then gasified in a surface gasifier unit to produce syngas. Underground coal gasification is a combined extraction and conversion process.

Experimentation on UCG was first performed in 1912 by Sir William Ramsay in England [4]. The experiments demonstrated the potential of UCG, but Ramsay's work was interrupted by the First World War. After the war, further UCG research did not continue, since coal was relatively inexpensive and available through conventional mining techniques in Western Europe [4]. The former Soviet Union was the first to begin considerable research and development programs with respect to large-scale UCG systems in the 1930s [15,16]. UCG technology was first utilized within commercial operations by the former Soviet Union for heating and power generation applications, which has continued to implement these systems for over 50 years [17]. Even though UCG has the appearance of being commercially mature, the technology from the former Soviet Union has been gaining interest only recently, with a rapid increase in the number of pilot plants throughout the rest of the world over the last decade. There are many commercial projects entering pilot plant phase and undergoing study, in Australia, New Zealand, the USA, India, Pakistan, Canada, and Italy. National research programs are being reconsidered in the USA and the UK, after preliminary systems failed to reach commercial maturity. As of 2008, the number of UCG trials includes 200 in the former Soviet Union, 33 in the USA, and approximately 40 distributed between South Africa, China, Australia, Canada, New Zealand, India, Pakistan, and Europe [17,18].

UCG Concepts and Technology

UCG is similar to surface gasification [19], with syngas produced through the same chemical reactions [12]. The main difference is that surface gasification occurs in a manufactured reactor whereas the reactor for a UCG system is a natural geological formation containing unmined coal [19, 20]. UCG also has similarities to in situ combustion processes applied in heavy-oil recovery and oil shale retorting, with such common operational parameters as roof/floor stability, seam continuity and permeability, and ground water influx [19, 21].

The basic UCG concept is illustrated in Figure 1. UCG involves an arrangement of injection and production wells drilled into coal seams. The coal is ignited and compressed gasification agents are fed into the coal seam through injection wells which triggers and controls an in situ sub-stoichiometric combustion process, producing syngas [22]. Syngas is extracted using production wells and is processed for use [17]. Suitable gasifying agents include air, oxygen, steam/air, and steam/oxygen [23]. The main difference between using oxygen and atmospheric air is that utilization of oxygen increases the heating value of the syngas [24], but producing pure oxygen requires additional energy and resources.

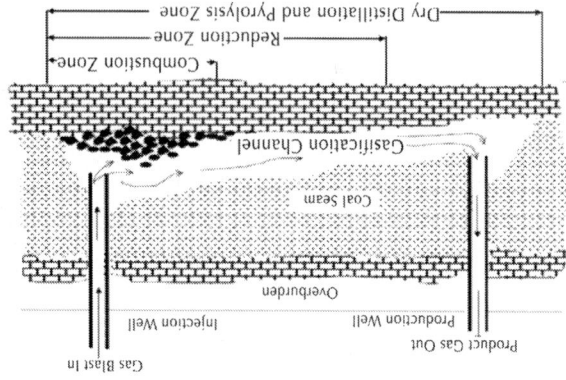

Figure 1: Schematic of in situ underground coal gasification process (modified from [21]).

Coal ignition is initiated through the use of an electric coil or gas firing near the face of the coal seam. Continuous oxidant flow through the injection well allows for gasification to be sustained [1]. The temperature of the gasification process is maintained through varying the oxidant flow to the reactor [23]. In UCG systems, the temperature of the coal face can reach temperatures in excess of 1,500 K [24, 25].

Various chemical reactions, temperatures, pressures, and gas compositions exist at different locations within a UCG gasifier. The gasification channel is normally divided into three zones: oxidization, reduction, and dry distillation and pyrolysis [26, 27]. In the oxidization zone, multiphase chemical reactions occur involving the oxygen in the gasification agents and the carbon in the coal. The highest temperatures in the gasifier occur in the oxidation zone, due to the large release of energy during the initial reactions [28]. The following reactions occur in the oxidation zone:

$$C + O_2 \rightarrow CO_2 + 393.8 \text{ kJ} \tag{1}$$

$$2C + O_2 \rightarrow 2CO + 231.4 \text{ kJ} \tag{2}$$

$$2CO + O_2 \rightarrow 2CO_2 + 571.2 \text{ kJ}. \tag{3}$$

In the reduction zone, the main reactions involve the reduction of H_2O (g) and CO_2 into H_2 and CO at high temperatures within the oxidation zone [26]. The following endothermic reactions occur in the reduction zone [21, 28]:

$$C + CO_2 \rightarrow 2CO - 162.4 \text{ kJ} \tag{4}$$

$$C + H_2O(g) \rightarrow CO + H_2 - 131.5 \text{ kJ}. \tag{5}$$

Under the catalytic action of coal ash and metallic oxides, a methanation reaction occurs:

$$C + 2H_2 \rightarrow CH_4 + 74.9 \text{ kJ}. \tag{6}$$

The energy terms, within the above equations, represent the amount of energy released or consumed during each reaction with the stoichiometric coefficients in equations representing moles. Equations 1 to 6 are taken from [21] and [28]. The endothermic reactions in the reduction zone decrease the temperature in the gasification channel to below that required for the reduction reactions.

Within the distillation (pyrolysis) zone, the coal seam is decomposed into multiple volatiles including H_2O, CO_2, CO, C_2H_6, CH_4, H_2, tar, and char [21,28]. At the exit of the gasification channel, the volatile composition of the syngas consists mostly of CO, H_2, and CH_4. The UCG process can also have other products, including H_2S, As, Hg, Pb, and ash [27, 29, 30]. The composition of syngas is highly dependent on the gasification agent, air injection method, and coal composition [31, 32]. During operation, the three gasification zones move along the direction of the air flow, ensuring continuous gasification reactions [21]. A distinguishing feature of UCG, compared to surface gasification, is that drying, pyrolysis, and char gasification occur simultaneously within the coal [26].

By-products of the UCG process pose an environmental hazard to the local surroundings through leaching of organic and inorganic materials into groundwater. Environmental data were first made available after later trials in the USA, including Hanna and Hoe Creek UCG trials, for which groundwater contamination monitoring was conducted before, during, and after gasification. The results illustrated that at shallow depths, UCG can pose a significant risk to groundwater in adjacent strata [30].

Groundwater pollution around UCG zones is mainly caused by one of the following mechanisms: dispersion and penetration of the pyrolysis products of the coal seam to the surrounding rock layers, the emission and dispersion of high contaminants with gas products after gasification, and migration of residue by leaching and penetration of groundwater [30]. In addition, the escaped gases such as carbon dioxide, ammonia, and sulfide may change the pH value of the local strata if dissolved.

The entire process is confined to the space of the coal seam and is sealed from the surface by natural geological formations or man-made barriers; the coal seam and strata serve, to some extent, as a natural groundwater cleaning system. In general, systems have active pressure control, in which the cavity pressure is held in equilibrium or below that of the surrounding strata [17, 30]. The pressure difference induces flow into the reactor space, which inhibits gasification products from leaking away from the cavity [33, 34].

The quality of the product gas is influenced by several parameters - such as the pressure inside the coal seam, coal properties, feed conditions, kinetics, and heat and mass transport within the coal seam - and the product of the UCG process is a multi-compound, high-temperature, and high-pressure syngas [1]. When the syngas reaches the surface, it is cleaned and undesired by-products are removed from the product stream [24]. Removal techniques are similar to those used with surface gasifiers. Once the by-products are removed, they can be disposed of safely, or used for other chemical processes [16]. The degree of cleaning required is dependent on the use of the syngas; syngas is cleaned either to meet the specification for input into a gas turbine (for electricity generation) or to be of sufficient purity for use as a chemical feedstock for conversion to synthetic fuels [20].

Methods for UCG. Two standard methods of preparing a coal seam for gasification have been utilized successfully: shaft and shaftless. The method implemented is dependent on parameters such as the natural permeability of the coal seam, geochemistry of the coal, seam thickness, depth, width and inclination, proximity of urban developments, and the amount of mining desired [34].

- *Shaft UCG methods*: Shaft methods use coal mine galleries and shafts to transport gasification reagents and products, which sometimes entail the creation of shafts and the drilling of large-diameter openings through underground labor [34]. The shaft method was the first technique utilized within UCG systems. Currently, the shaft method is only employed in closed coal mines due to economic and safety reasons [34]. The following are examples of common UCG shaft methods:

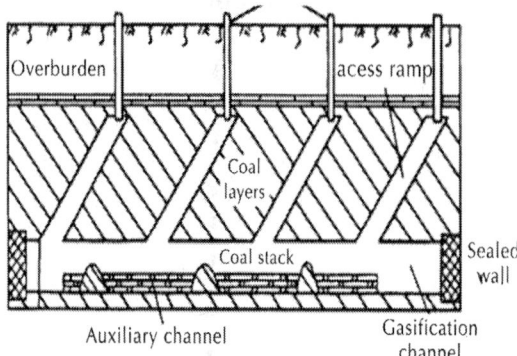

Figure 2: Schematic of the structure of a LLT underground gasifier (modified from [28]).

- *Chamber or warehouse method*: This method utilizes constructed underground galleries with brick walls separating coal panels. Gasification agents are supplied to a previously ignited coal face on one side of the wall, and the syngas is removed from a gallery on the other side. The chamber method strongly relies on the natural permeability of the coal seam to allow for sufficient oxidant flow through the system. The syngas composition may vary during operation, and the gas production rates are often low. To improve system output, coal seams are often outfitted with explosives for rubblization prior to the reaction zone [21].

- *Borehole producer method*: For this method, parallel underground galleries are created within a coal seam with sufficient distance between them. The galleries are connected by drilling boreholes from one gallery to the other [34]. Remote electric ignition of the coal in each borehole is used to initiate the gasification process. This method is designed to gasify considerably flat-lying seams. Some variations exist where linking of the galleries is accomplished through hydraulic and electric linking [21, 34].

- *Stream method*: This method is designed for sharply inclined coal beds. Parallel pitched galleries following the contour of the coal seam are constructed and are connected at the bottom of the seam by a horizontal gallery also known as a fire-drift. To initiate gasification, fire is introduced within the horizontal gallery. The hot coal face moves up the seam slope with oxidant fed through one inclined gallery and syngas leaving through the other. The main advantage of this method is that the ash and roof material drop down to fill the void space created during the process, which prevents suffocating the gasification process at the coal front [21].

- *LLT gasification method*: This method utilizes mined tunnels or constructed roadways to connect the injection well to the production well [4]. Typical long and large tunnel (LLT) systems consist of a gasification channel, two auxiliary holes, and two auxiliary tunnels (Figure 2). The auxiliary holes are arranged between the injection and production wells and are used as malfunction holes for the injection of air and water vapor, or to discharge gas for added gasifier control. LLT also includes an auxiliary tunnel constructed of bricks, which is an auxiliary installation for air injection that prevents blockage in the gasification channel. The mined tunnels are isolated by sealing walls to prevent leakage of combustible gases from the gasifier [35]. The location and height of the oxidant injection points and gas outlet points can be adjusted, allowing for two-dimensional control of oxidant injection and gas production [28].

- *Shaftless UCG methods*: Recently, most of the focus of research has been on the shaftless methods, which employ directional drilling techniques [6]. The preparation of a reactor for the directional drilling technique consists of the creation of dedicated in-seam boreholes for oxidant injection and product collection using drilling and completion technology that has been adapted from oil and gas production.

With shaftless methods, all preparation and operational processes are carried out through a series of boreholes drilled from the surface into a coal seam and do not require underground labor. Preparation of a shaftless reactor consists of the creation of dedicated in-seam boreholes for oxidant injection and product collection using drilling and completion technology that has been adapted from oil and gas production [34]; the approach generally includes drilling inlet and outlet boreholes into a coal seam, increasing the coal permeability between the inlet and outlet boreholes, igniting the coal seam, introducing an oxidant for gasification, and extraction of the product gas from the outlet well[21]. Currently, there are two main classifications of shaftless UCG methods: linked vertical well (LVW) and controlled retractable injection point (CRIP).

- *LVW method*: The LVW method is one of the oldest methods for UCG and is derived from technology developed in the former Soviet Union [16]. Vertical wells are drilled into a coal seam, and internal pathways in the coal are utilized to direct the oxidant and product gas flow from the inlet to the outlet borehole. Internal pathways can be naturally occurring or constructed [35]. In its simplest form, the LVW method has inlet and outlet borehole locations that are static for the life of the system. During operation, the coal face migrates and it is found that system control, performance, and syngas quality are affected negatively as the distance from the coal face to the oxidant injection point increases [4]; this factor greatly reduces the feasibility of simple LVW systems.

A more advanced LVW approach involves a series of dedicated injection boreholes located along the length of a coal seam [21]. Over the life of a UCG reactor, the coal face, being gasified, travels as localized coal is exhausted [4]. Having multiple boreholes for injection allows for improved static operating conditions. A more complex variation of the LVW method also exists where multiple inlet and outlet boreholes are drilled into a coal seam, forming inlet and outlet borehole pairs. Parallel inlet and outlet manifolds are connected to the boreholes to provide a path for oxidant and syngas flows, respectively. Coal between each pair of inlet and outlet boreholes forms a zone. When the coal in a zone has been exhausted, new boreholes are drilled in a location of fresh coal, forming new zones [21]. Low-rank coals, such as lignites, have considerable natural permeability and can be exploited for UCG without the need for linking technologies. However, high-rank coals, such as anthracites, are far less permeable, making the gas production rate more limited if UCG is employed [35]. For the use of high-rank coals in UCG, a method of linking must be employed to increase the permeability and fracture the coal seam [36]. The boreholes in traditional LVW gasifiers are linked by special methods including forward combustion, reverse combustion, fire linkage, electric linkage, hydrofracturing, and directional drilling to create sizable gasification channels [35, 37].

Controlled retractable injection point. Over the span of a coal seam, the geometry may change, resulting in variable UCG operation and system performance [38]. In the past, this problem was solved by having multiple injection and/or production wells so that static operating conditions could be accomplished through moving the gasifier zones to fresh coal [16]. CRIP offers an alternative approach where the vertical injection well is not moved, but the injection point is moved within the coal seam to fresh coal when necessary [39].

The CRIP method relies on a combination of conventional drilling and directional drilling to access the coal seam and physically form a link between the injection and production wells, without the use of linking technologies utilized in LVW methods [38]. A vertical section of injection well is drilled to a predetermined depth, after which directional drilling is used to expand the hole and drill along the bottom of the coal seam creating a horizontal injection well [40]. At the end of the injection well, a gasification cavity is initiated in a horizontal section of the coal seam, creating a localized reactor. The CRIP system utilizes a burner attached to retractable coiled tubing which is used to ignite the coal [39]. The burner burns through the borehole casing to ignite the coal. The ignition point can be moved to any desired location along the horizontal injection well for the creation of a new gasification cavity after a deteriorating reactor has been deserted [38]. Typically, the injection point is retracted using a gas burner, which burns a section of the liner at a desired location [39]. In this manner, accurate control of the gasification process can be obtained. This UCG method has gained popularity in Europe and the USA, but the use of the CRIP method for UCG is fairly new and currently has not become commonly employed [4].

UCG with CO_2 Capture and Storage

All fossil fuels emit CO_2 when combusted. Currently, coal has the highest CO_2 emissions, per unit energy produced, of the fossil fuels used in combustion [13, 41]. To maintain and expand the use of coal, implementation of CCS technologies is becoming imperative.

CO_2 capture can be performed in three main fashions: pre-combustion, post-combustion, and oxy-firing [42]. A broad range of technology options are available for capturing CO_2 including physical absorption, chemical absorption, membrane separation, and cryogenic separation [42, 43]. Within UCG, the syngas compositions, temperatures, and pressures of production streams at the exit of a production well are comparable to those of surface gasifiers, which allow similar methods of CO_2 capture. Due to similarities, it is believed that UCG syngas could take advantage of separation using physical sorbents, within a pre-combustion arrangement, which has costs comparable to capture technologies commonly utilized in integrated gasification combined cycles [4, 12]. Post-combustion methods are also applicable and would be directly comparable in terms of cost and performance to typical post-combustion systems utilized in power plants. Oxy-firing options are possible for UCG as well, and within a power generating scenario, an air separation unit can generate O_2 streams for injection into the UCG and for use in an oxy-fired plant utilizing the syngas [12].

The spatial coincidence of geological carbon storage (GCS) options with UCG opportunities suggests that designers could colocate and combine UCG and GCS systems with high potential for effective CO_2 storage [18]. In general, these storage options would be the same for conventional carbon sequestration operations, including saline formations and mature oil and gas fields [44]. For UCG-CCS utilizing conventional sequestration options, there could exist common interests in site characterization and monitoring between UCG and CCS projects, where work performed during the design and implementation of one project could be used within the other. Coordinating UCG and CCS designs would improve economics for both projects.

If UCG and CCS are coupled, there is an attractive carbon management scheme associated, where most of the expected CO_2 emissions are sequestered back into a coal seam void that has been recently created by spent subsurface reactors through existing injection and production wells [13, 21]. When voids are created, they typically collapse, similar to voids produced during longwall coal mining, leaving zones of artificial breccias with high permeability. Suitable containment zones prevent vertical flow of CO_2 to the surface, where storage locations are isolated from the surface by low-permeability strata (known as seals or caprocks, often shales or evaporites) [4, 45]. For a spent UCG system to accommodate CO_2 storage, the void must be at depths below approximately 800 to 1,000 m [44-46]. These depths are required so that supercritical pressures and temperatures exist that allow the CO_2 density to be high enough (approximately 500 to 700 kg/m³) to limit the storage volume required [45].

The UCG-CCS approach, if successful, could offer an integrated energy recovery and CO_2 storage system, which exploits a new sequestration resource created during operation. A significant challenge with CCS is its large energy requirement [47], of which a considerable portion is consumed during CO_2 capture and compression [48]. The pressure after compression is generally high enough to allow for a reduction in pressure during transport while allowing the fluid to be in a liquid state [9]. If CO_2 storage is accommodated in spent UCG reactors, CO_2 transport and compression requirements decline. CO_2 transport accounts for 5% to 15% of a conventional CCS financial budget, which can be lowered with a self-contained UCG-CCS project, through reduced piping and shipping requirements associated with long-distance transport [18]. A large portion of the budget for a CCS project is allotted for CO_2 storage, typically 10% to 30%, most of which is used for geological and geophysical studies and drilling injection wells [18, 48]. These tasks are commonly completed during UCG construction and would not need to be repeated for the implementation of CCS, thus reducing system cost relative to conventional storage methods [18].

As of 2009, it remains unclear if CCS using UCG-produced voids is viable [44]. Until recently, this alternative has received little attention, and there remains substantial scientific uncertainty associated with the technological challenges and environmental risks of storing CO_2 in this manner [13, 44]. For full-scale commercialization, extensive research and development is needed to alleviate the uncertainties. Currently, CO_2 sequestration is under development internationally by such organizations as the Intergovernmental Panel on Climate Change and Carbon Sequestration Leadership Forum [13].

CONCLUSIONS

Although the earth is an abundant source of coal, a significant amount is currently unrecoverable. With the introduction of UCG, recoverable coal reserves can be expanded by possibly a couple hundred years. Coal is likely to remain used in many countries, increasing the needs for new technologies that permit more environmentally benign extraction and utilization. Wide-scale use of UCG is such a technology option, with the syngas it produces usable as a fuel. Fossil fuels typically utilized in power production could then be used for other purposes, which would result in large reductions in their consumption rates. UCG offers a coal extraction and conversion method in a single process that avoids many of the challenges associated with conventional mining practices. UCG has a high potential for integration with CCS using conventional methods utilized in power production due to similarities with surface gasifier units. UCG also has the potential to store CO_2 within voids created during its operation, which reduces the need for transport and storage site identification. In essence, UCG could provide a cost-effective, near-zero-carbon, energy source through the use of a self-contained system with a closed carbon loop.

AUTHORS' CONTRIBUTIONS

SS carried out the review and drafted the manuscript. BR and MR co-supervised the investigation, conceived of the study, reviewed the manuscript, and helped draft parts. All authors read and approved the final manuscript.

ACKNOWLEDGMENTS

The authors acknowledge the financial support of the Natural Sciences and Engineering Research Council of Canada.

REFERENCES

1.   Daggupati, S, Mandapati, RN, Mahajani, SM, Ganesh, A, Pal, AK, Sharma, RK, Aghalayam, P: Compartment modeling for flow characterization of underground coal gasification cavity. Ind. Eng. Chem. Res. 50, 277–290 (2011).

2.   Tanaka, N(e): World Energy Outlook 2009, International Energy Agency, Paris (2009)

3.   Energy Information Administration: International Energy Outlook 2010, Department of Energy, U.S. Government, Washington (2010)

4.   Roddy, DJ, Younger, PL: Underground coal gasification with CCS: a pathway to decarbonising industry. Energy Environ. Sci. 3, 400–407 (2010).

5.   Ediger, VS, Hosgor, E, Surmeli, AN, Tatlidil, H: Fossil fuel sustainability index: an application of resource management. Energy Policy. 35, 2969–2977 (2007).

6.   Hammond, GP: Energy, environment and sustainable development: a UK perspective. Transactions of the Institution of Chemical Engineers, Part B. 78, 304–323 (2000)

7.   Aleklett, K, Hook, M, Jakobsson, K, Lardelli, M, Snowden, S,

Soderbergh, B: The peak of the oil age–analyzing the world oil production reference scenario in World Energy Outlook 2008. Energy Policy. 38, 1398–1414 (2010).

8. World Energy, C: Survey of Energy Resources 2007, London, World Energy Council (2007)

9. Ghose, MK, Paul, B: Underground coal gasification: a neglected option. Int. J. Environ. Stud.64(6), 777–783 (2007).

10. Couch, G: Underground Coal Gasification, IEA Clean Coal Centre, London (2009)

11. Shackley, S, Mander, S, Reiche, A: Public perceptions of underground coal gasification in the United Kingdom. Energy Policy. 34, 3423–3433 (2006).

12. Burton, E, Friedmann, J, Upadhye, R: Best Practices in Underground Coal Gasification, Lawrence Livermore National Laboratory, Livermore (2006)

13. Khadse, A, Qayyumi, M, Mahajani, S, Aghalayam, P: Underground coal gasification: a new clean coal utilization technique for India. Energy. 32, 2061–2071 (2007).

14. Breault, RW: Gasification processes old and new: a basic review of the major technologies. Energies. 3, 216–240 (2010).

15. Gregg, DW, Hill, RW, Olness, DU: An Overview of the Soviet Effort in Underground Gasification of Coal, Lawrence Livermore National Laboratory, Livermore (1976)

16. Shafirovich, E, Varma, A: Underground coal gasification: a brief review of current status. Ind. Eng. Chem. Res. 48, 7865–7875 (2009).

17. Van der Riet, M: Underground coal gasification. Proceedings of the SAIEE Generation Conference, Eskom College, Midrand (19 Feb 2008)

18. Roddy, D, Gonzalez, G: Underground coal gasification (UCG) with carbon capture and storage (CCS). In: Hester RE, Harrison RM (eds.) Issues in Environmental Science and Technology, pp. 102–125. Royal Society of Chemistry, Cambridge (2010)

19. Pana, C: Review of Underground Coal Gasification with Reference to Alberta's Potential, Energy Resources Conservation Board, Edmonton (2009)

20. Walker, L: Underground coal gasification: a clean coal technology ready for development. The Australian Coal Review. 8, 19–21 (1999)

21. Lee, S, Speight, JG, Loyalka, SK: Handbook of Alternative Fuel Technologies, CRC, Boca Raton (2007)

22. Kempka, T, Plötz, ML, Schlüter, R, Hamann, J, Deowan, SA, Azzam, R: Carbon dioxide utilisation for carbamide production by application of the coupled UCG-urea process. Energy Procedia. 4, 2200–2205 (2011)

23. Perkins, G, Sahajwallaa, V: Modelling of heat and mass transport phenomena and chemical reaction in underground coal gasification. Chem. Eng. Res. Des.. 85(3), 329–343 (2007).

24. Perkins, G, Sahajwalla, V: Steady-state model for estimating gas production from underground coal gasification. Energy Fuel. 22, 3902–3914 (2008).

25. Peng, FF, Lee, IC, Yang, RYK: Reactivities of in situ and ex situ coal chars during gasification in steam at 1000-1400 °C. Fuel Process. Technol. 41, 233–251 (1995).

26. Perkins, G, Sahajwallaa, V: A mathematical model for the chemical reaction of a semi-infinite block of coal in underground coal gasification. Energy Fuel. 19, 1679–1692 (2005).

27. Yang, LH, Pang, XL, Liu, SQ, Chen, F: Temperature and gas pressure features in the temperature-control blasting underground coal gasification. Energy Sources: Part A. 32, 1737–1746 (2010)

28. Yang, L, Liang, J, Yu, L: Clean coal technology—study on the pilot project experiment of underground coal gasification. Energy. 28, 1445–1460 (2003).

29. Liu, S, Wang, Y, Yu, L, Oakey, J: Thermodynamic equilibrium study of trace element transformation during underground

coal gasification. Fuel Process. Technol. 87, 209–215 (2006).

30. Shu-qin, L, Jing-gang, L, Mei, M, Dong-lin, D: Groundwater pollution from underground coal gasification. J. China Univ. Mining & Technol. 17(4), 0467–0472 (2007).

31. Sta czyk, K, Howaniec, N, Smoli ski, A, wiadrowski, J, Kapusta, K, Wiatowski, M, Grabowski, J, Rogut, J: Gasification of lignite and hard coal with air and oxygen enriched air in a pilot scale ex situ reactor for underground gasification. Fuel. 90, 1953–1962 (2011).

32. Prabu, V, Jayanti, S: Integration of underground coal gasification with a solid oxide fuel cell system for clean coal utilization. Int. Journal of Hydrogen Energy. 37, 1677–1688 (2012).

33. Yang, LH: A review of the factors influencing the physicochemical characteristics of underground coal gasification. Energy Sources, Part A. 30(11), 1038–1049 (2008).

34. Wiatowski, M, Sta czyk, K, wi drowski, J, Kapusta, K, Cybulski, K, Krause, E, Grabowski, J, Rogut, J, Howaniec, N, Smoli ski, A: Semi-technical underground coal gasification (UCG) using the shaft method in Experimental Mine "Barbara". Fuel. 99, 170–179 (2012)

35. Liang, J, Liu, S, Yu, L: Trial study on underground coal gasification of abandoned coal resource. In: Xie H, Golosinki TS (eds.) Proceedings of the ‹99 International Symposium on Mining Science and Technology, Beijing, August 1999. Mining Science and Technology 99, pp. 271–275. A.A. Balkema, Rotterdam (1999).

36. Blinderman, MS, Klimenko, AY: Theory of reverse combustion linking. Combustion and Flame.150, 232–245 (2007).

37. Blinderman, MS, Saulov, DN, Klimenko, AY: Forward and reverse combustion linking in underground coal gasification. Energy. 33, 446–454 (2008).

38. Nourozieh, H, Kariznovi, M, Chen, Z, Abedi, J: Simulation study of underground coal gasification in Alberta reservoirs:

geological structure and process modeling. Energy Fuel. 24, 3540–3550 (2010).

39. Klimenko, AY: Early ideas in underground coal gasification and their evolution. Energies. 2, 456–476 (2009).

40. Wang, GX, Wang, ZT, Feng, B, Rudolph, V, Jiao, JL: Semi-industrial tests on enhanced underground coal gasification at Zhong-Liang-Shan coal mine. Asia-Pac. J. Chem. Eng.. 4, 771–779 (2009).

41. Nag, B, Parikh, J: Indicators of carbon emission intensity from commercial energy use in India. Energy Economics. 22, 441–461 (2000).

42. Göttlicher, G, Pruschek, R: Comparison of CO2 removal systems for fossil-fuelled power plant processes. Energy Convers. Mgmt. 38, 173–178 (1997).

43. Ho, MT, Allinson, G, Wiley, DE: Comparison of CO_2 separation options for geo-sequestration: are membranes competitive? Desalination. 192, 288–295 (2006).

44. Friedmann, SJ, Upadhye, R, Kong, FM: Prospects for underground coal gasification in carbon-constrained world. Energy Procedia. 1(1), 4551–4557 (2009).

45. Orr, FM: Onshore geologic storage of CO_2. Science. 325(5948), 1656–1658 (2009).

46. Budzianowski, WM: Value-added carbon management technologies for low CO2 intensive carbon-based energy vectors. Energy. 41, 280–297 (2012).

47. Steinberg, M: Fossil fuel decarbonization technology for mitigating global warming. International Journal of Hydrogen Energy. 24, 771–777 (1999).

48. Gibbins, J, Chalmers, H: Carbon capture and storage. Energy Policy. 36, 4317–4322 (2008).

Chapter 3

Al$_2$O$_3$-based Nanofluids: A Review

Veeranna Sridhara[1] and
Lakshmi Narayan Satapathy[2]

[1]New Horizon College of Engineering, Bangalore, India
[2]Ceramic Technological Institute, BHEL, Malleswaram Complex, Bangalore 560012, India

ABSTRACT

Ultrahigh performance cooling is one of the important needs of many industries. However, low thermal conductivity is a primary limitation in developing energy-efficient heat transfer fluids that are required for cooling purposes. Nanofluids are engineered by suspending nanoparticles with average sizes below 100 nm in heat transfer fluids such as water, oil, diesel, ethylene glycol, etc.

Innovative heat transfer fluids are produced by suspending metallic or nonmetallic nanometer-sized solid particles. Experiments have shown that nanofluids have substantial higher thermal conductivities compared to the base fluids. These suspended nanoparticles can change the transport and thermal properties of the base fluid. As can be seen from the literature, extensive research has been carried out in alumina-water and CuO-water systems besides few reports in Cu-water-, TiO_2-, zirconia-, diamond-, SiC-, Fe_3O_4-, Ag-, Au-, and CNT-based systems. The aim of this review is to summarize recent developments in research on the stability of nanofluids, enhancement of thermal conductivities, viscosity, and heat transfer characteristics of alumina (Al_2O_3)-based nanofluids. The Al_2O_3 nanoparticles varied in the range of 13 to 302 nm to prepare nanofluids, and the observed enhancement in the thermal conductivity is 2% to 36%.

INTRODUCTION

Conventional fluids, such as water, engine oil, and ethylene glycol are normally used as heat transfer fluids. Although various techniques are applied to enhance the heat transfer, the low heat transfer performance of these conventional fluids obstructs the performance enhancement and the compactness of heat exchangers. The use of solid particles as an additive suspended into the base fluid is technique for the heat transfer enhancement. Improving the thermal conductivity is the key idea to improve the heat transfer characteristics of conventional fluids. Since a solid metal has a larger thermal conductivity than a base fluid, suspending metallic solid fine particles into the base fluid is expected to improve the thermal conductivity of that fluid. The enhancement of thermal conductivity of conventional fluids by the suspension of solid particles, such as millimeter- or micrometer-sized particles, has been well-known for many years [1]. However, they have not been of interest for practical applications due to problems such as sedimentation leading to increased pressure drop in the flow channel. The recent advance in material technology has made it possible to produce innovative heat transfer fluids by suspending nanometer-sized particles in base fluids which can change the transport and thermal properties of the base fluid.

Nanofluids are solid-liquid composite materials consisting of solid nanoparticles or nanofibers with sizes typically of 1 to 100 nm suspended in liquid. The nanofluid is not a simple liquid-solid mixture; the most important criterion of nanofluid is agglomerate-free stable suspension for long durations without causing any chemical changes in the base fluid. This can be achieved by minimizing the density between solids and liquids or by increasing the viscosity of the liquid; by using nanometer-sized particles and by preventing particles from agglomeration, the settling of particles can be avoided. Nanofluids have attracted great interest recently because of reports of enhanced thermal properties [2-6]. Extensive research has been carried out on alumina-water- and CuO-water-based systems besides few reports in Cu-water, carbon nanotubes water systems.

This article aims at an overview of the concept of "alumina-based Nanofluids" followed by an account on the detailed research activities carried out around the world. The review will focus mainly on engineering application parameters, such as thermal conductivity and viscosity etc., without giving much emphasis on the theoretical aspects.

Preparation of Nanofluids

There are two fundamental methods to obtain nanofluids:

- Single-step direct evaporation method: In this method, the direct evaporation and condensation of the nanoparticulate materials in the base liquid are obtained to produce stable nanofluids.

- Two-step method: In this method, first the nanoparticles are obtained by different methods and then are dispersed into the base liquid.

Thermal Conductivity Measurement Techniques

Thermal conductivity is an important parameter in enhancing the heat transfer performance of a base fluid. Since the thermal conductivity of solid metals is higher than that of fluids, the suspended particles are expected to increase the thermal conductivity and heat transfer performance. Many researchers have reported experimental studies on the thermal conductivity of nanofluids. The temperature oscillation method [7], the steady-state parallel plate method [8], and transient hot-wire method [1] have been employed to measure the thermal conductivity of nanofluids. However, the transient hot-wire method has been extensively used by many researchers [9-12]. A detailed review on different techniques for measurement of thermal conductivity of nanofluids is available in the literature [5].

Experimental Results on Thermal Conductivity of Al_2O_3-based Nanofluids

Alumina (Al_2O_3) is the most common nanoparticle used by many researchers in their experimental works. Many efforts have been made to study the thermal conductivity of nanofluids. The summary of experimental studies on the thermal conductivity of Al_2O_3-based nanofluids are given in Table 1. Generally, thermal conductivity of the nanofluids increases with increasing volume fraction of nanoparticles; with decreasing particle size, the shape of particles can also influence the thermal conductivity of nanofluids, temperature, Brownian motion of the particle, interfacial layer, and with the additives.

Table 1: The selective summary of the thermal conductivity enhancement in Al$_2$O$_3$-based nanofluids

Author (year)	Base fluid	Concentration	Particle size (nm)	Enhancement ratio	Method/parameters
Masuda et al.[15]	Water (31.85°C) Water (46.85°C) Water (66.85°C)	1.3 to 4.3	13	1.1092 to 1.324 1.10 to 1.296 1.092 to 1.262	Two-step method Temperature effect
Lee et al. [1]	Water Ethylene	1.0 to 4.30 1.0 to 5.0	38.4	1.03 to 1.10 1.03 to 1.18	Two-step method
Wang et al. [8]	Water Ethylene glycol Engine oil Pump oil	3.0 to 5.50 5.0 to 8.0 2.25 to 7.40 5.00 to 7.10	28	1.11 to 1.16 1.25 to 1.41 1.05 to 1.30 1.13 to 1.20	Two-step method
Eastman et al.[18]	Ethylene glycol	1.00 to 5.00	35		Two-step method
Xie et al. [16]	Water Ethylene glycol Ethylene glycol Ethylene glycol Ethylene glycol Pump oil	1.80 to 5.001.80 to 5.00 1.80 to 5.00 1.80 to 5.00 5.00	60.4 15 26 60.4 302 60.4	1.07 to 1.21 1.06 to 1.17 1.06 to 1.18 1.10 to 1.30 1.08 to 1.25 1.39	Two-step method Solid crystalline Phase effect Morphology effect pH effect
Xie et al. [16]	Water Ethylene glycol Pump oil Glycerol	5.0 5.0 5.0 5.0	60.4	1.39 1.23 1.29 1.38	Two-step method Base fluid effect
Das et al. [7]	Water (21°C) Water (36°C) Water (51°C)	1.00 to 4.00	38.4	1.02 to 1.09	Two-step method Temperature effect

Wen and Ding[24]	Water + sodium dodecyl benezene sulfonate	0.19 to 1.59	42	1.01 to 1.10	Two-step method
Li and Peterson [13]	Water (27.5°C) Water (32.5°C) Water (34.7°C)	2.00 to 10.00	36	1.08 to 1.11 1.15 to 1.22 1.18 to 1.29	Two-step method Temperature effect
Beck et al.[11]	EG (27°C)	1.00 to 4.00	20	1.015 to 1.14	Two-step method
Hwang et al.[17]	Water	0.3 to 1.0	48	1.013 to 1.04	Two-step method
Timofeeva et al. [14]	Water EG	5.0 5.0 5.0 5.0	11 20 40 All sizes	1.08 1.07 1.10 1.13	Two-step method Temperature effect (296°C to 333°C)
Lee et al. [19]	Water	0.01 to 03	35	1.005 to 1.02	Two step
Murshed et al.[20]	Water EG CTAB	1.0 0.5 1.0	80 150 80 80	1.03 to 1.12 1.02 to 1.10 1.03 to 1.09 1.06 to 1.12	Two-step method Temperature range (21°C to 60°C)
Choi et al. [12]	Transformer oil + oleic acid	0.5 to 4.0	13	1.05 to 1.20	Two-step method
Oh et al. [23]	Water EG	1.0 to 4.0 1.0 to 4.0	45 45	1.044 to 1.133 1.019 to 1.097	Two-step method
Kole et al. [25]	Car engine coolant	3.5	50	1.1041 to 1.1125	Two-step method Temperature effect (30°C to 80°C)

Sridhara and Satapathy *Nanoscale Research Letters* 2011 6:456, doi: 10.1186/1556-276X-6-456

Effect of Volume Fraction of Nanoparticles on Thermal Conductivity of Al$_2$O$_3$-based Nanofluids

The effect of volume concentration on Al$_2$O$_3$-based nanofluids is shown in Figure 1. The researchers used different sizes of Al$_2$O$_3$ nanoparticles at different temperatures in water and ethylene glycol with particle volume concentration mostly less than 5% with few exceptions [13,14]. The maximum enhancement in thermal conductivity observed for 4 vol.% load in the case of water-based nanofluid was 32% [15] and in the case of ethylene glycol-based nanofluid was 30% [16], respectively. Hwang et al. [17] observed a 4% enhancement in thermal conductivity at 1 vol.% concentration; the observed enhancement was more compared to other researchers at same the volume fraction of solids [1,11,18]. Lee et al. [19] observed a 2% enhancement at a lower volume percent for 35-nm-sized Al$_2$O$_3$ particles. In the case of Li and Peterson [13], the thermal conductivity enhancement was decreased as concentration increased from 6% to 10%, but in the case of Timofeeva et al. [14], the thermal conductivity was increased as concentration increased from 2% to 10% even though the particle size was almost the same in both the cases.

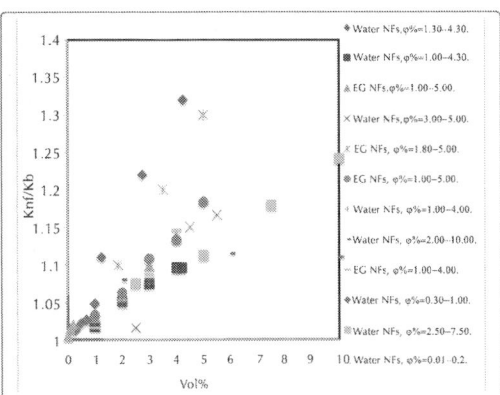

Figure 1: Effect of concentration on thermal conductivity of Al$_2$O$_3$-based nanofluids. Water NFs, ϕ% = 1.30-4.30 [15]; water NFs, ϕ% = 1.00-4.30

[1]; EG NFs, ϕ% = 1.00-5.00 [1]; water NFs, ϕ% = 3.00-5.00[8]; EG NFs, ϕ% = 1.80-5.00 [16]; EG NFs, ϕ% = 1.00-5.00 [18]; water NFs, ϕ% = 1.00-4.00 [7]; water NFs, ϕ% = 2.00-10.00 [13]; EG NFs, ϕ% = 1.00-4.00 [11]; water NFs, ϕ% = 0.30-1.00 [17]; water NFs, ϕ% = 2.50-7.50 [14]; water NFs, ϕ% = 0.01-0.20 [19].

Effect of Particle Size on Thermal Conductivity of Al_2O_3 Based Nanofluids

Figure 2 demonstrates the effect of particle size on thermal conductivity of Al_2O_3-based nanofluids; the particles used were in the range of 13 to 150 nm. Alumina, 38.4 nm, in water resulted in thermal conductivity enhancement in the range of 2% to 10% in two studies [1,7] but up to 21% in another study [16]. The thermal conductivity enhancement for the nanofluids with 28 nm [8] particles was lying in between that of 38.4 and 60.4 nm, which cannot be explained. Murshed et al. [20] observed higher enhancement with 80- and 150-nm-sized particles at 1 vol.% compared to the nanofluids with 2.5 vol.% of 28-nm particles in ethylene glycol-based Al_2O_3 nanofluids as shown in Figure 3. The authors have demonstrated that 80-nm particles showed higher thermal conductivity enhancement at 1 vol.% compared to similar data reported earlier [1,11,14]. Xie et al. [16] used 15- and 60.4-nm-sized particles, observed higher thermal conductivity enhancement for larger nanoparticles in ethylene glycol-based nanofluids. The results cited here do not correlate the size effect of nanoparticles in thermal conductivity enhancement. More research is required to understand this size effect.

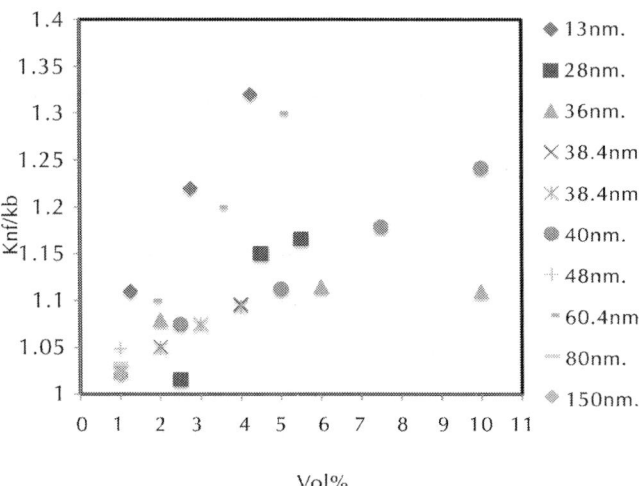

Figure 2: Effect of particle size on thermal conductivity of water-based Al$_2$O$_3$ nanofluids. 13 nm [15], 28 nm [8], 36 nm [13], 38.4 nm[1], 38.4 nm [7], 40 nm [14], 48 nm [17], 60.4 nm [16], 80 nm [20], 150 nm [20].

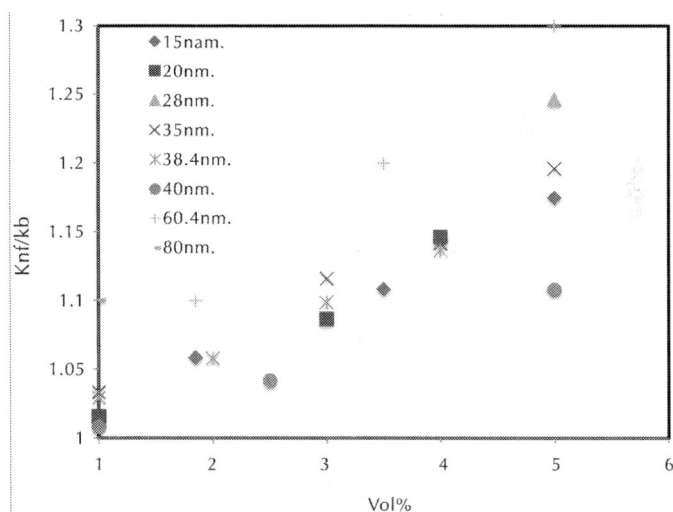

Figure 3: Effect of particle size on thermal conductivity of ethylene gly-col-based Al$_2$O$_3$ nanofluids. 15 nm [16], 20 nm [11], 28 nm [8], 35 nm [18], 38.4 nm [1], 40 nm [14], 60.4 nm [16], 80 nm [20].

Effect of Base Fluids on Thermal Conductivity of Al₂o₃-based Nanofluids

The effect of base fluid on thermal conductivity is shown in Figure 4. The result in Figure 4demonstrates that the thermal conductivity enhancement is least for the water-based nanofluids compared with other nanofluids. This result is encouraging because heat transfer enhancement is often most needed when poorer heat transfer fluids are involved. The enhancement in the case of PO is 38% at 5 vol.% compared to that of 20% at 4 vol.% TO in contrast to 10.8% enhancement with the same volume fraction of nanoparticles in water [21]. Figure 4 thus categorically indicated that the thermal conductivity enhancement for the poorer heat transfer fluids is good compared to the fluids with better thermal conductivity such as water.

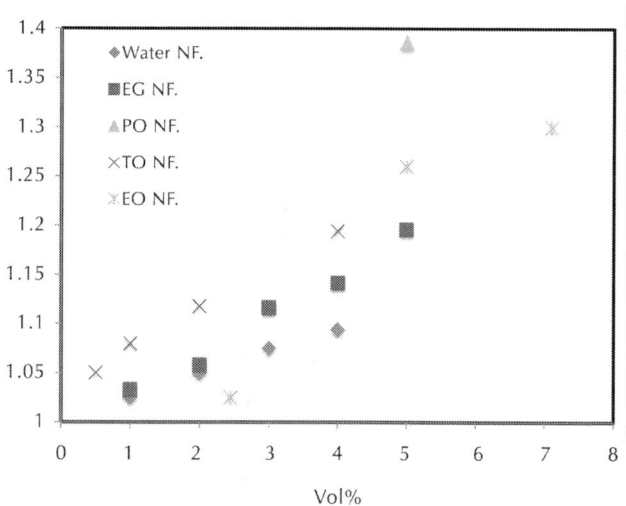

Figure 4: Effect of base fluids on thermal conductivity of Al₂O₃-based nanofluids. Water NF [7], EG NF [18], PO NF [16], TO NF [12], EO NF [8].

Effect of Preparation Method on Thermal Conductivity of Al$_2$o$_3$-based Nanofluids

Thermal conductivities of the nanoparticle fluid mixture were first reported by Masuda et al. [15]. The mean diameter of the particles used in their experiments was 13 nm, and the particles dispersed in water by using a high-speed shearing dispenser ≈ 20,000 rpm. The authors reported a 32.4% increase in thermal conductivity for the volume fraction of 4.3 vol.% against 20% for 3 vol.% nano alumina. However, the experiment was carried out at a higher room temperature of approximately 32°C, which is higher than most other researchers' reported data at room temperature ranging from 21°C to 28°C. Further, the authors used a high-speed dispenser with addition of HCl and NaOH to the fluids so that electrostatic repulsive forces among the particles kept the powder well dispersed. Lee et al. [1] dispersed 38.4-nm-sized Al$_2$O$_3$ nanoparticles in water and ethylene glycol by using polyethylene container and shaken thoroughly to ensure a homogeneous suspension for producing stable suspension. The authors observed an increase of only 10% at the 4.3 vol.% and 8% for the 3% load. The same enhancement was observed by Das et al. [7] for the particle size of 38.4 nm and for the particle load between 1% and 4%. Wang et al. [8] dispersed 28-nm-sized Al$_2$O$_3$ nanoparticles in different base fluids and prepared nanofluids by mechanical blending, coating particles with polymers and filtration method. The thermal conductivity enhancement was 16% for 5.5 vol.% and 12% for 3% volume faction. In the case of Xie et al. [16], the researchers used 60.4-nm-sized Al$_2$O$_3$ dispersed in water and prepared stable solution by adjusting pH. The nanoparticles are de-agglomerated by using an ultrasonic disrupter after mixing with a base fluid and were homogenized by using magnetic force agitation. The enhancement observed was 21% for 5% volume fraction and 14% at 3.2% volume fraction.

Figure 5 shows that the enhancement in the case of Xie et al. [16] is more compared to others even though they used lesser particles and in the case of Wang et al. [8] shows lesser enhancement at 2.5 vol.% compared to Das et al. [7] and Lee et al. [1]. The same method of synthesis in the latter two cases demonstrated similar enhancement ratios. These results demonstrate that a stable dispersion can be achieved by many different methods, but the thermal conductivity enhancement is dependent on the preparation methods. These results need further studies since the stability of such fluids in the long run has not been studied and the data reported here is immediately after obtaining the nanofluid.

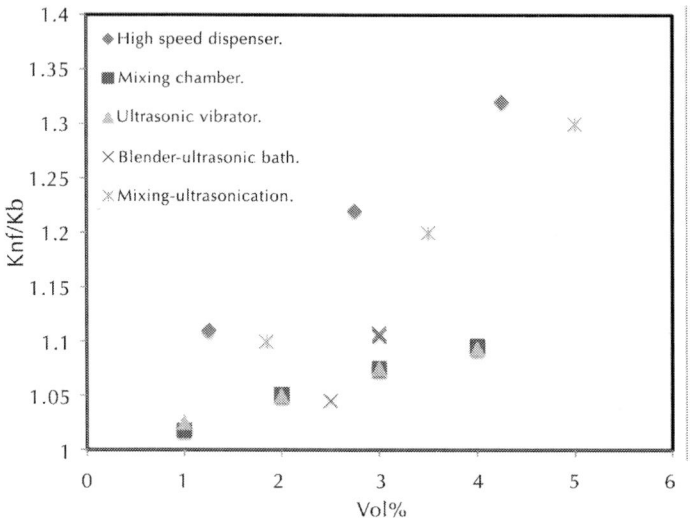

Figure 5: Effect of preparation techniques on thermal conductivity of Al_2O_3-based nanofluids. High-speed disperser [15], mixing chamber [1], ultrasonic vibrator [7], blender-ultrasonic bath [8], mixing-ultrasonication [16].

Effect of Temperature on Thermal Conductivity of Al_2O_3-based Nanofluids

The thermal conductivity of nanofluids is temperature sensitive compared to that of base fluids. The effect of temperature on water-based Al_2O_3 nanofluids is shown in Figure 6. Different groups measured thermal conductivity at different temperatures. Das et al. [7] varied temperatures in the range of 21°C to 51°C demonstrating an enhancement of 2% to 10.8% for the particle load of 2 vol.% and observed thermal conductivity enhancement of 9.4% as compared to 24.3% for 4 vol.% solids. The authors suggested that strong temperature dependence of nanofluid thermal conductivity is due to the motion of the particles. The larger sized particles used by Murshed et al. [20] resulted in enhancement similar to that reported earlier [7] indicating that the enhancement is due to the intensification of the Brownian motion of the nanoparticles by addition of a surfactant and the application of temperature. The general trend in Figure 6 of increased thermal conductivity enhancement with increased temperature is not in line with a very early report of Masuda et al. [15].

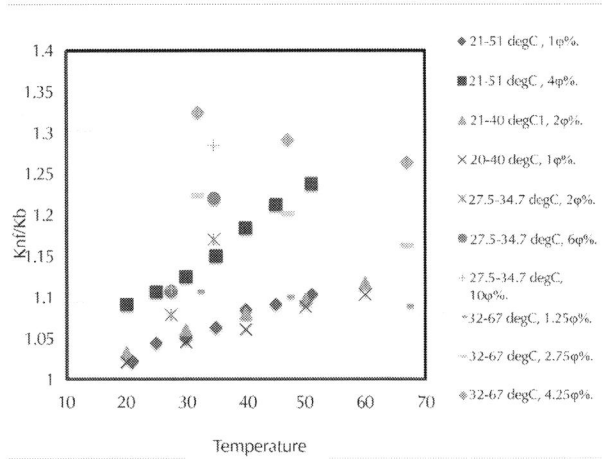

Figure 6: Effect of temperature on thermal conductivity of Al$_2$O$_3$-based nanofluids. 21°C to 51°C, 1 ϕ% [7]; 21°C to 51°C, 4 ϕ% [7]; 20°C to 40°C, 2 ϕ% [20]; 20°C to 40°C, 1 ϕ% [20]; 27.5°C to 34.7°C, 2 ϕ% [13]; 27.5°C to 34.7°C, 6 ϕ% [13]; 27.5°C to 34.7°C, 10 ϕ% [13]; 52°C to 67°C, 1.25 ϕ% [15]; 52°C to 67°C, 2.75 ϕ% [15]; 52°C to 67°C, 4.25 ϕ% [15].

In Figure 7, the same trend is observed for ethylene glycol-based nanofluids. Both Murshed et al.[20] and Beck et al. [22] observed higher conductivity enhancement for the suspensions containing surfactants, though particle size of solids were different in both cases. Recently, Beck *et al.* [22] measured thermal conductivity of the ethylene glycol-based nanofluids in the range of 296 to 400 K and showed that thermal conductivity behavior of nanofluids is related the behavior of the base fluid, and they suggested that temperature dependence of nanofluids is due mostly to the base fluids.

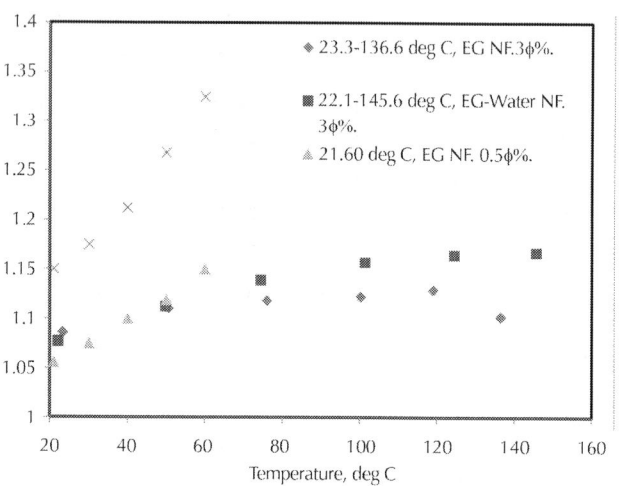

Figure 7: Effect of temperature on EG and EG + H$_2$O-based Al$_2$O$_3$ nanofluids. 23.3 to 136.6, EG NF, 3 ϕ% [22]; 222.1 to 145.6, EG-water NF, 3 ϕ% [22]; 21°C to 60°C, EG NF, 0.5 ϕ% [20]; 21°C to 60°C, EG NF, 1.0 ϕ% [20].

These results of temperature dependence of thermal conductivity enhancement in nanofluids based on alumina during the last 15 years since 1993 is confusing and hence needs thorough analysis. The data must be interpreted in conjunction with the base fluid behavior, particle size, and surfactant effect.

Thermal Conductivity of Al$_2$O$_3$ Nanofluids Measured by Different Techniques

Figure 8 shows the thermal conductivity measurement of Al$_2$O$_3$ water-based nanofluid measured by different techniques. A trend shows that thermal conductivity increased with the increase in volume fraction. The thermal conductivity data in the case of Oh et al. [23] were in well agreement with that reported by Wang et al. [8] which, however, was higher than the results of Lee et al. [1] and Das et al. [7] for similar nanofluids but measured by different techniques. The reason for this discrepancy during the measurement may be due to the sedimentation and aggregation of nanoparticles, particle diameter, and nanofluid preparation. In comparing the thermal conductivity measurement techniques, the steady state parallel plate method seems to be least affected by the particle sedimentation since the thickness of the loaded sample fluid is less than 1 mm. The transient hot-wire method can be affected by the sedimentation of the nanofluids. Non-homogeneous nanoparticle concentration in the direction of gravity can give rise to temperature gradient within the vertical hot wire, which may be a source of measurement errors. This is also true for the temperature oscillation technique [23]. It is not clear how these techniques will behave for a stable nanofluid which does not at all sediment during the measurement. Therefore, it is essential to produce nanofluids which can be stable for long periods of time without any noticeable sedimentation.

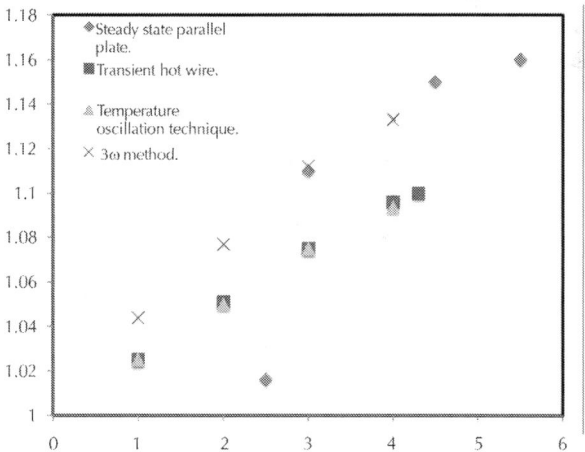

Figure 8: Thermal conductivity of Al$_2$O$_3$ nanofluids measured by different

techniques. Steady-state parallel plate [8]; transient hot-wire method [1]; temperature oscillation technique [7]; 3 method [23].

Effect pH on Thermal Conductivity of Al$_2$O$_3$ Water-based Nanofluids

Xie *et al.* [16] prepared various suspensions containing Al$_2$O$_3$ nanoparticles with specific surface areas in a range of 5 to 124 m^2/g, and their thermal conductivities were measured using a transient hot-wire method at a pH range of 2 to 11.5. It was noted that the nanoparticle suspensions, containing a small amount of Al$_2$O$_3$, have substantially higher thermal conductivity than the base fluid, with the enhancement increasing with the volume fraction of Al$_2$O$_3$. The enhanced thermal conductivity increases with an increase in the difference between the pH value of aqueous suspension and the isoelectric point of the Al$_2$O$_3$ particle. The enhancement observed for 60.4-nm-sized particle between 1.8 and 5 vol.% is 7% to 21%. The effect of pH on thermal conductivity of water-based Al$_2$O$_3$ nanofluids is shown in Figure 9.

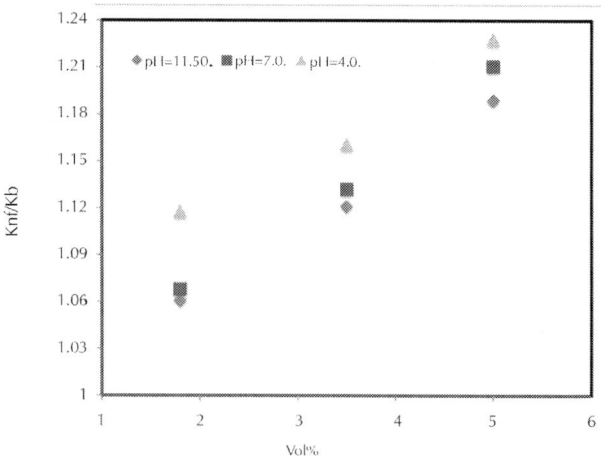

Figure 9: Effect of pH on thermal conductivity of water-based Al$_2$O$_3$nano-fluids. pH = 11.5 [16]; pH = 7.0 [16]; pH = 4.0 [16].

Effect of Surface Active Agents on Thermal Conductivity of Water-based Al$_2$o$_3$ Nanofluids

Figure 10 compares the thermal conductivity enhancement of Al$_2$O$_3$ nanofluids with and without a surfactant. Wen *et al.* [24] used 42-nm-sized Al$_2$O$_3$ nanoparticles and dispersed them in water using sodium dodecyl benzene sulfonate (SDBS) as surfactant; the enhancement observed was 10% for 1.59 vol.% which is comparable with the data reported earlier [1,7,18]. Recently, Kole *et al.* [25] dispersed < 50-nm-sized Al$_2$O$_3$ using oleic acid as surfactant in a car engine coolant and observed 10.41% enhancement for 3.5 vol.%. The authors have demonstrated the stability of such fluids for more than 80 days with thermal conductivity enhancement of 13% and 12% for ethylene glycol-based Al$_2$O$_3$ nanofluids at 5 vol.% solid loading. As shown in Figure 10, the additives will enhance the thermal conductivity of the nanofluids and give good stability, but the question which is unresolved is the contribution of thermal conductivity enhancement from the surfactant effect to the overall enhancement of the nanofluids.

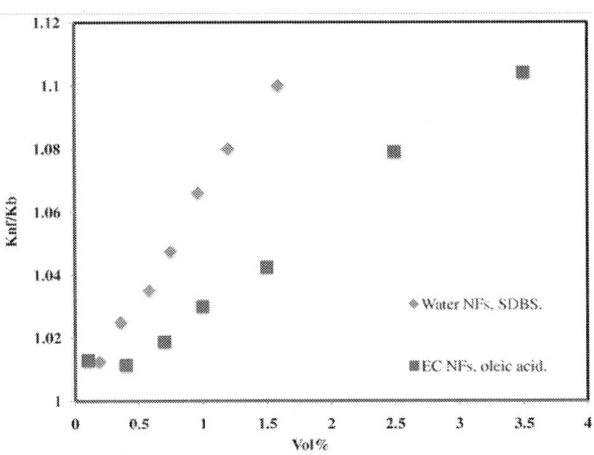

Figure 10: Effect of additives on thermal conductivity of Al$_2$O$_3$ nanofluids. Water NFs, SDBS [24]; EC NFs, oleic acid [25].

Experimental Results on Viscosity of Al_2O_3-based Nanofluids

Compared with the experimental studies on thermal conductivity of nanofluids, there are limited rheological studies reported in the literature. In one study [8], the Al_2O_3-water mixture showed a viscosity increase between 20% and 30% for 3 vol.% Al_2O_3 solution compared to that of water alone. The results by Das *et al.* [26] on the viscosity of alumina-water nanofluids against shear rate demonstrated an increase of viscosity with increased particle concentrations indicating strong possibility that nanofluid may be non-Newtonian. Further investigations are, however, required to define the viscosity models of nanofluids.

In another study, a two-step method was used to produce Al_2O_3-water nanofluids with low concentrations of Al_2O_3 nanoparticles from 0.01 to 0.3 vol.% without any surfactant [19] and measured viscosity at the temperature range from 21°C to 39°C. Experimental results showed that the effective viscosities of the dilute Al_2O_3-water nanofluids significantly decreases with increasing temperature and slightly increases with increasing volume fraction. The measured viscosity of the Al_2O_3-water nanofluids is nonlinear with the Al_2O_3 nanoparticle volume concentration. The nonlinear viscosity behavior occurs at very low particle concentrations far below 2 vol.%. Nonlinear behavior implies that there is particle-particle interactions which invalidate the Einstein equation developed for dilute suspensions. The result is similar in another experiment [27], wherein, the viscosity increased by 83.4% at a volume fraction of 0.05 (5 vol.%). The viscosity study on Al_2O_3-water nanofluids with 36- and 47-nm, and CuO-water nanofluid with 29-nm average particle size was reported by Nguyen*et al.* [28] for particle volume fraction ranging from 1% to 9.4% and for temperatures varying from room temperature to approximately 75°C. The hysteresis behavior for 36-nm particle size and four particle volume concentrations indicated drastic changes with heating of samples beyond a critical temperature. On cooling after being heated beyond a critical temperature, a hysteresis phenomenon can occur. It is very interesting to note that hysteresis is predominant only in fluids with higher nanoparticle concentration.

The measured viscosities of Al$_2$O$_3$ (80 nm) and deionized water (DIW)-based nanofluids were also found to increase by nearly 82% for the maximum volumetric loading of 5% nanoparticles [20]. A similar increment (86%) of the effective viscosity of Al$_2$O$_3$ (28 nm)/ distilled water-based nanofluids was also observed by Wang et al. [8] for the same volume fraction of 0.05. The reasons for the differences could be due to the difference in the size of the particle clusters, differences in the dispersion techniques, and the use of a surfactant similar to that reported earlier for thermal conductivity data. At lower concentrations, the change in relative viscosity over temperature was minimal. Xie et al. [29] also demonstrated that the viscosity of the nanoparticle suspension is much larger than the corresponding value predicted by the theoretical formula. The enhancements ratio of the viscosity of ethylene glycol (EG)-based suspensions are smaller than those of water-based suspensions, indicating the significant influence of the base fluid on the viscosity of the fluid-nanoparticle mixtures. The recent report by Kole et al. [25] of alumina in engine oil demonstrated that there is a transition from Newtonian characteristics for the base fluid to non-Newtonian behavior with increasing content of Al$_2$O$_3$ in the engine coolant. The data also show that the viscosity increases with an increase in concentration and decreases with an increase in temperature.

The analysis of limited data indicated that an optimization is required for the solid loading in nanofluids so that the viscosity rise is not high for the application and at the same time there is significant enhancement in the thermal conductivity of nanofluids. More studies are required in this direction.

Heat Transfer Characteristics of Al$_2$o$_3$-based Nanofluids

While heat transfer aspects of suspensions are important in applications in general, the aspect of natural convection in multiphase emulsions becomes more critical during storage and special phenomena such as melting of clathrate, which is used for storing coldness by releasing latent heat, separates out as organic liquid and an emulsion of hydrofluorocarbon dispersed in water [4].

Pak and Cho [30] studied the heat transfer enhancement in a circular tube, using γ-Al_2O_3 and TiO_2 nanoparticle fluid mixtures as the flowing medium. They observed an increase in the Nusselt number with the increasing volume fraction and Reynolds number. Putra *et al.* [31] studied the natural convection of nanofluids inside horizontal cylinder heated from one end and cooled from the other. An apparently paradoxical behavior of heat transfer deterioration was observed in the experimental study. The nature of this deterioration and its dependence on parameters such as particle concentration, material of the particles, and geometry of the containing cavity was investigated. The fluid characters are distinct from that of common slurries.

Heris *et al.* [32] dispersed CuO and Al_2O_3 oxide nanoparticles in water as base fluid in different concentrations, and the laminar flow convective heat transfer through circular tube with constant wall temperature boundary condition were examined. The experimental results obtained for CuO-water and Al_2O_3-water nanofluids indicate that heat transfer coefficient ratios for nanofluid to homogeneous model in low concentrations are close to each other, but by increasing the volume fraction, higher heat transfer enhancement for Al_2O_3/water was observed. The same authors worked on laminar flow forced convection heat transfer of Al_2O_3/water nanofluid inside a circular tube with constant wall temperature [33] and measured the Nusselt numbers for different nanoparticle concentrations as well as various Peclet and Reynolds numbers. Experimental results emphasized the enhancement of heat transfer due to the presence of nanoparticles in the fluid. Heat transfer coefficient increased by increasing the concentration of nanoparticles in nanofluid.

The turbulent convective heat transfer behavior of alumina (Al_2O_3) and zirconia (ZrO_2) nanoparticle dispersions in water is investigated experimentally in a flow loop with a horizontal tube test section at various flow rates (9,000 < Re < 63,000) [34], The experimental data were compared to predictions made using the traditional single-phase convective heat transfer and viscous pressure loss correlations for fully developed turbulent flow, Dittus-Boelter, and Blasius/MacAdams, respectively. It was shown that if the measured temperature- and loading-dependent thermal conductivities and viscosities of the nanofluids are used in calculating the Reynolds, Prandtl, and Nusselt numbers, the existing correlations accurately reproduce the convective heat transfer and viscous pressure loss behavior in tubes. Therefore, no abnormal heat transfer enhancement was observed in this study.

Xuan and Li [35] conducted an experiment to investigate convective heat transfer and flow features of the nanofluid in a tube. Both the convective heat transfer coefficient and friction factor of the sample nanofluids for the turbulent flow were measured, respectively. The effects of such factors as the volume fraction of suspended nanoparticles and the Reynolds number on the heat transfer and flow features are discussed in detail. Wen and Ding [24] reported an experimental work on the convective heat transfer of nanofluids, made of γ-Al$_2$O$_3$ nanoparticles and DIW, flowing through a copper tube in the laminar flow regime. The results showed considerable enhancement of convective heat transfer using the nanofluids; the enhancement was particularly significant in the entrance region, and was much higher than that solely due to the enhancement on thermal conduction. The possible reasons for the enhancement are migration of nanoparticles and the resulting disturbance of the boundary layer.

You et al. [36] measured the critical heat flux (CHF) in the pool boiling of Al$_2$O$_3$-water nanofluids. They discovered an unprecedented phenomenon: a threefold increase in CHF over that of pure water. The average size of departing bubbles increased, and the bubble frequency decreased significantly in nanofluids when compared with those in pure water. Bang et al. [37] studied boiling heat transfer characteristics of Al$_2$O$_3$-based nanofluids. Pool boiling heat transfer coefficients and phenomena of nanofluids are compared with those of pure water, which are acquired on a smooth horizontal flat surface (roughness of a few tens of nanometers). The experimental results showed that these nanofluids have poor heat transfer performance compared to pure water in natural convection and nucleate boiling. On the other hand, CHF has been enhanced in not only horizontal but also vertical pool boiling. This is related to a change of surface characteristics by the deposition of nanoparticles.

Experimental study conducted by Das *et al.* [38] on pool boiling in water-Al_2O_3 nanofluids on horizontal tubes of small diameter revealed that the deterioration in performance in boiling is less in narrow tubes compared to that in large industrial tubes which makes it less susceptible to local overheating in convective application. Recently, Farajollahi *et al.* [39] measured heat transfer characteristics of -Al_2O_3/water and TiO_2/water nanofluids in a shell and tube heat exchanger under turbulent flow condition. It was reported that by adding nanoparticles to the base fluid, significant enhancement of heat transfer characteristics was observed. For both nanofluids, two different optimum nanoparticle concentrations exist. A comparison of the heat transfer behavior of two nanofluids indicates that at a certain Peclet number, the heat transfer characteristics of TiO_2/water nanofluid at its optimum nanoparticle concentration are greater than those of -Al_2O_3/water nanofluid while -Al_2O_3/water nanofluid possesses better heat transfer behavior at higher nanoparticle concentrations. In another recent study, it was demonstrated that the alumina nanofluids significantly improved the thermal performance of an oscillating heat pipe [40], with an optimal mass fraction of 0.9 wt.% for maximal heat transfer enhancement. Compared with pure water, the maximal thermal resistance was decreased by 0.14°C/W (or 32.5%) when the power input was 58.8 W at 70% filling ratio and 0.9% mass fraction. The authors observed that the nanoparticle settlement mainly took place at the evaporator. The change of surface condition at the evaporator due to nanoparticle settlement was found to be the major reason for the enhanced thermal performance of the alumina nanofluid-charged oscillating heat pipe.

Recently, Ho *et al.* [41], conducted an experiment on natural convection of heat transfer of a nanoluid in vertical square enclosures of different sizes, in the solid loading range of 0.1 to 4 vol.% noted the Rayleigh's number varying in the range of 6.21 × 10^3 to 2.56 × 10^8. The experimental result for the average heat transfer rate across the three enclosures appeared generally consistent with the assessment based on the changes in thermophysical properties of the nanofluid formulated, showing systematic heat transfer degradation for the nanofluid containing nanoparticles volume fraction ≥2 vol.% over the entire range of the Rayleigh's number considered. The nanofluid containing 0.1 vol.%, a heat transfer enhancement of 18% compared with that of water, was found to arise in the largest enclosure at sufficiently high Rayleigh's number. The authors suggested that such enhancement is not only due to the relative changes in thermophysical properties of the nanofluid containing low particle fraction, other factors may come into play.

Although addition of local losses (orificing) may suppress instabilities, however, it is accompanied by a significant flow reduction which is detrimental to the natural circulation heat removal capability. Nayak et al. [42] demonstrated experimentally, with Al$_2$O$_3$ nanofluids, not only the flow instabilities are suppressed but also the natural circulation flow rate is enhanced. The increase in steady natural circulation flow rate due to addition of nanoparticles is found to be a function of its concentration in water. The flow instabilities are found to occur with water alone only during a sudden power addition from cold condition, step increase in power, and step decrease in power (step back conditions). With a small concentration of Al$_2$O$_3$ nanofluids, these instabilities are found to be suppressed significantly.

The heat transfer studies on alumina-based nanofluids can give rise to the possibility of their use in actual applications. However, cost of such fluids is a major concern vis-à-vis the stability duration of such fluids in ideal condition. Further, the effect of acids and bases or surfactants used for stabilization of nanoparticles in actual applications needs to be studied in detail.

Applications of Alumina-based Nanofluids

A nanofluid can be used to cool automobile engines and welding equipment and to cool high heat flux devices such as high-power microwave tubes and high-power laser diode arrays. A nanofluid coolant could flow through tiny passages in MEMS too to improve its efficiency. The measurement of nanofluid CHF in a forced convection loop is useful for nuclear applications. If nanofluids improve chiller efficiency by 1%, a savings of 320 billion kWh of electricity or an equivalent 5.5 million barrels of oil per year would be realized in the USA alone [43]. Nanofluids find a potential for use in deep drilling application. A nanofluid can also be used for increasing the dielectric strength and life of the transformer oil by dispersing nanodiamond particles.

Nguyen *et al.* [44] experimentally investigated the behavior and heat transfer enhancement of an Al_2O_3 nanoparticle-water mixture, flowing inside a closed system that is destined for the cooling of microprocessors or other electronic components. Experimental data, obtained for turbulent flow regime, have clearly shown that the inclusion of nanoparticles into distilled water has produced a considerable enhancement of the cooling block convective heat transfer coefficient. For a particular nanofluid with 6.8% particle volume concentration, heat transfer coefficient has been found to increase as much as 40% compared to that of the base fluid. It has also been found that an increase of particle concentration has produced a clear decrease of the heated component temperature. Experimental data also showed that a nanofluid with a 36-nm particle provides higher heat transfer coefficients than a 47-nm particle size. In another experiment, You *et al.* [36] measured the enhancement of the CHF in pool boiling from a flat square heater immersed in alumina-based water nanofluid in a concentration range of 0 to 0.05 g/l. The test results showed that the enhancement of CHF was drastic when nanofluid was used as a cooling liquid instead of pure water. It was concluded that the increase in CHF levels present the possibility of raising chip power in electronic components or simplifying cooling requirements for space applications. Tzeng *et al.* [45] dispersed

CuO and Al$_2$O$_3$ nanoparticles and antifoam, respectively into cooling engine oil for the cooling of automotive transmission. The experimental platform was a four-wheel drive transmission vehicle. It adopts advanced rotary blade coupling (RBC), where a high local temperature occurs easily at high rotating speeds. Therefore, it is imperative to improve the heat transfer efficiency. The experiment measures the temperature distribution of the RBC exterior at four different rotating speeds (400, 800, 1,200, and 1,600 rpm), simulating the conditions of a real car at different rotating speeds and investigating the optimum possible compositions of a nanofluid for higher heat transfer performance. Kulkarni et al.[46] used Al$_2$O$_3$ nanofluid as a coolant in a diesel electric generator. Specific heat measurements of aluminum oxide nanofluid with various particle concentrations were studied and showed that applying nanofluids resulted in a reduction of cogeneration efficiency. This is due to the decrease in specific heat, which influences the waste heat recovery from the engine. However, it was found that the efficiency of waste heat recovery heat exchanger increased for nanofluid due to its superior convective heat transfer coefficient.

Recently, Wu et al. [47] observed the potential of Al$_2$O$_3$-H$_2$O nanofluids as a new phase change material for the thermal energy storage of cooling systems. The thermal response test shows the addition of Al$_2$O$_3$ nanoparticles remarkably decreases the super cooling degree of water, advances the beginning freezing time, and reduces the total freezing time. The infrared imaging photographs suggest that the freezing rate of nanofluids is enhanced and by only adding 0.2 wt.% Al$_2$O$_3$ nanoparticles, the total freezing time of Al$_2$O$_3$-H$_2$O nanofluids can be reduced by 20.5%. Transformer cooling is important to the Navy as well as the power generation industry with the objective of reducing transformer size and weight. The ever growing demand for greater electricity production can lead to the necessity of replacing and/or upgrading transformers on a large scale and at a high cost. A potential alternative in many cases is the replacement of conventional transformer oil with a nanofluid. Such retrofits can represent considerable cost savings. It has been demonstrated that the heat transfer properties of transformer oils can be significantly improved by using nanoparticle additives [48]. The above experimental results demonstrated that alumina-based nanofluids have a significant potential in applications. However, large volumes of nanofluid experiments are lacking in literature. Further, most of the applications are limited to closed loop configuration. A need thus arises to test such fluids with suitable modifications in open loop applications.

Summary

Alumina-based nanofluids are important because they can be used in numerous applications involving heat transfer and other applications. Most of the Al_2O_3-based nanofluids are prepared by using an ultrasonic vibrator which is not stable for a longer time.Researchers therefore had concentrated on preparing stable nanofluids by using different surfactants, optimizing pH, temperature for different nanofluids, and by surface modification of the particles. The thermal conductivity enhancement observed for Al_2O_3 nanofluid by different researchers is not consistent; the reason for this enhancement is not clear in the available literature. Very few literatures are available on the enhancement of thermal conductivity due to surface area, acidic or basic media, and due to the shape factor. The nanofluids prepared with acidic and basic media may not be useful for the heat transfer application, since it may cause adverse effects on the heat transfer properties. The effect of temperature observed by different authors demonstrates different degrees of enhancement for the same volume fraction. The technique for the measurement of thermal conductivity may also alter the values. The effect of temperature on thermal conductivity at lower volume fractions, which has been measured up to 400 K, has been reported . No work has yet been reported with experiments dealing with the measurement of thermal conductivity at low (sub-zero)-range temperatures. The behavior of the thermal conductivity at low temperatures are yet to be found out and can point a new direction in this field of research. Very few reports described the effect of temperature on the viscosity at a higher volume fraction and at a higher temperature observing hysteresis phenomenon. The heating phase beyond the critical temperature may become more viscous which indicates a rather drastic alteration of the nanofluid rheological properties leading to hysteresis. Viscosity has raised a serious concern regarding the use of nanofluids for enhancing heat transfer mobility. Researchers can concentrate on the effect of temperature and the hysteresis behavior for Al$_2$O$_3$ nanofluids and can try to increase the temperature withstanding capacity of the Al$_2$O$_3$ nanofluids. From the observed results, it is clearly seen that nanofluids have a greater potential for heat transfer enhancement and are suitable for application in practical heat transfer processes.

The review has summarized the basics of nanofluid, its preparation methods, and the factors affecting the thermal conductivity enhancement in the Al$_2$O$_3$-based nanofluid. It has also identified the areas which require more research for better understanding. The enhancement of thermal conductivity of base fluid will be a definite requirement in the future to improve the thermal efficiency of different systems.

AUTHORS' CONTRIBUTIONS

SV compiled the studies conducted on thermal conductivity, viscosity, heat transfer and pool boiling phenomena of alumina based nanofluids, compared and analysed the results.

LNS involved in the conceptualizing the manuscript and revising it critically for improving through technical concepts. Both the authors read and approved the final manuscript.

REFERENCES

1.   Lee S, Choi SUS, Li S, Eastman JA: Measuring thermal conductivity of fluids containing oxide nanoparticles.*ASME J Heat Transfer* 1999, 121:280-89.

2.   Wang XQ, Mujumdar AS: A review on nanofluids - part I:

theoretical and numerical investigations.*Braz J Chem Eng* 2008, 25:613-630.

3.   Singh AK: Thermal conductivity of nanofluids.*Defence Sci J* 2008, 58:600-607.

4.   Sridhara V, Gowrishankar BS, Snehalatha C, Satapathy LN: Nanofluids--a new promising fluid for cooling.*Trans Ind Ceram Soc* 2009, 68:1-17.

5.   Paul G, Chopkar M, Manna IA, Das PK: Techniques for measuring the thermal conductivity of nanofluids: a review. *Renew Sust Energ Rev* 2010, 14:1913-1924.

6.   Godson L, Raja B, Mohan Lal D, Wongwises S: Enhancement of heat transfer using nanofluids--an overview.*Renew Sust Energ Rev* 2010, 14:629-641.

7.   Das SK, Putra N, Thiesen P, Roetzel W: Temperature dependence of thermal conductivity enhancement for nanofluids.*ASME J Heat Transfer* 2003, 125:567-574.

8.   Wang X, Xu X, Choi SUS: Thermal conductivity of nanoparticle-fluid mixture.*J Thermophys Heat Trans* 1999, 13:474-480.

9.   Xuan Y, Li Q: Heat transfer enhancement of nanofluids.*Int J Heat Fluid Fl* 2000, 21:58-64.

10.   Murshed SMS, Leong KC, Yang C: Enhanced thermal conductivity of TiO_2--water based nanofluids.*Int J Therm Sci* 2005, 44:367-373.

11.   Beck MP, Sun T, Teja AS: The thermal conductivity of alumina nanoparticles dispersed in ethylene glycol.*Fluid Phase Equilib* 2007, 260:275-278.

12.   Choi C, Yoo HS, Oh JM: Preparation and heat transfer properties of nanoparticles-in-transformer oil dispersions as advanced energy-efficient coolants.*Curr Appl Phys* 2008, 8:710-712.

13.   Li CH, Peterson GP: Experimental investigation of temperature and volume fraction variations on the effective thermal conductivity nanoparticle suspensions (nanofluids).*J Appl Phys* 2006, 99:084314.

14. Timofeeva EV, Gavrilov AN, McCloskey JM, Tolmachev YV: Thermal conductivity and particle agglomeration in alumina nanofluids: experiment and theory.*Phys Rev E* 2007, 76:061203.

15. Masuda H, Ebata A, Teramae K, Hishinuma N: Alteration of thermal conductivity and viscosity of liquid by dispersing ultra-fine particles (dispersion of -Al$_2$O$_3$, SiO$_2$, and TiO$_2$ ultra-fine particles).*Netsu Bussei* 1993, 7:227-233.

16. Xie H, Wang J, Xi T, Liu Y, Ai F: Thermal conductivity enhancement of suspensions containing nanosized alumna particles.*J Appl Phys* 2002, 91:4568-4572.

17. Hwang D, Hong KS, Yang HS: Study of thermal conductivity nanofluids for the application of heat transfer fluids. *Thermochim Acta* 2007, 455:66-69.

18. Eastman JA, Choi SUS, Li S, Yu W, Thomson LJ: Anomalously increased effective thermal conductivities of ethylene glycol-based nanofluids containing copper nanoparticles.*Appl Phys Lett* 2001, 78:718-720.

19. Lee JH, Hwang KS, Jang SP, Lee BH, Kim JH, Choi SUS, Choi CJ: Effective viscosities and thermal conductivities of aqueous nanofluids containing low volume concentrations of Al$_2$O$_3$ nanoparticles.*Int J Heat Mass Trans* 2008, 51:2651-656.

20. Murshed SMS, Leong KC, Yang C: Invesitions of thermal conductivity and viscosity of nanofluids.*Int J Therm Sci* 2008, 47:560-568.

21. Yu W, France DM, Routbort JL, Choi SUS: Review and comparison of nanofluid thermal conductivity and heat transfer enhancements.*Heat Transfer Eng* 2008, 29:432-460.

22. Beck MP, Yuan Y, Warrier P, Teja AS: The thermal conductivity of alumina nanofluids in water, ethylene glycol, and ethylene glycol + water mixtures.*J Nanopart Res* 2010, 12:1469-1477.

23. Oh DW, Jain A, Eaton JK, Goodson KE, Lee JS: Thermal conductivity measurement and sedimentation detection of aluminum oxide nanofluids by using 3 method.*Int J Heat Fluid Fl* 2008, 29:1456-1461.

24. Wen D, Ding Y: Experimental investigation into convective heat transfer of nanofluids at the entrance region under laminar flow conditions.*Int J Heat Mass Trans* 2004, 47:5181-5188.

25. Kole M, Dey TK: Thermal conductivity and viscosity of Al_2O_3 nanofluid based on car engine coolant.*J Phys D Appl Phys* 2010, 43:315501.

26. Das SK, Putra N, Roetzel W: Pool boiling characteristics of nano-fluids.*Int J Heat Mass Trans* 2003, 46:851-862.

27. Liu MS, Lin MC, Huang IT, Wang CC: Enhancement of thermal conductivity with CuO for nanofluids.*Chem Eng Technol* 2006, 29:72-77.

28. Nguyen CT, Desgranges F, Roy G, Galanis N, Mare T, Boucher S, Angue Mintsa H: Viscosity data for Al_2O_3-water nanofluid--hysteresis: is heat transfer enhancement using nanofluids reliable?*Int J Heat Fluid Fl* 2007, 28:1492-506.

29. Xie H, Chen L, Wu Q: Measurements of the viscosity of suspensions (nanofluids) containing nanosized Al_2O_3 particles.*High Temp-High Press* 2008, 37:127-135.

30. Pak BC, Cho YI: Hydrodynamic and heat transfer study of dispersed fluids with submicron metallic oxide particles.*Exp Heat Tran* 1998, 11:151-170.

31. Putra N, Roetzel W, Das SK: Natural convection of nanofluids. *Heat Mass Transf* 2003, 39:775-784.

32. Heris SZ, Etemad SG, Esfahan MN: Experimental investigation of oxide nanofluids laminar flow convective heat transfer.*Int Commun Heat Mass Trans* 2006, 33:529-535.

33. Heris SZ, Esfahany MN, Etemad SG: Experimental investigation of convective heat transfer of Al_2O_3/water nanofluid in circular tube.*Int J Heat Fluid Fl* 2007, 28:203-210.

34. Williams W, Buongiorno J, Wen Hu L: Experimental investigation of turbulent convective heat transfer and pressure loss of alumina/water and zirconia/water nanoparticle colloids (Nanofluids) in horizontal tube.*J Heat Trans* 2008, 130:042412.

35. Xuan Y, Li Q: Investigation on convective heat transfer and flow features of nanofluid.*J Heat Trans* 2003, 125:151-155.

36. You SM, Kim JH, Kim KM: Effect of nanoparticles on critical heat flux of water in pool boiling heat transfer.*Appl Phys Lett* 2003, 83:3374-3376.

37. Bang IC, Chang SH: Boiling heat transfer performance and phenomena of Al$_2$O$_3$-water nano-fluids from a plain surface in a pool.*Int J Heat Mass Trans* 2005, 48:2407-2419.

38. Das SK, Putra N, Roetzel W: Pool boiling of nano-fluids on horizontal narrow tubes.*Int J Multiphase Fl* 2003, 29:1237-1247.

39. Farajollahi B, Etemad SG, Hojjat M: Heat transfer of nanofluids in a shell and tube heat exchanger.*Int J Heat Mass Trans* 2010, 53:12-17.

40. Qu J, Wu HY, Cheng P: Thermal performance of an oscillating heat pipe with Al$_2$O$_3$-water nanofluids.*Int Commun Heat Mass Trans* 2010, 37:111-115.

41. Ho CJ, Liu WK, Chang YS, Lin CC: Natural convection heat transfer of alumina-water nanofluid in vertical square enclosures: an experimental study.*Int J Therm Sci* 2010, 49:1345-1353.

42. Nayak AK, Gartia MR, Vijayan PK: Thermal-hydraulic characteristics of a single-phase natural circulation loop with water and Al$_2$O$_3$ nanofluids.*Nucl Eng Des* 2009, 239:526-540.

43. Nanofluids [http://www.uitonline.it/doc/NANO-FLUIDS.pdf] *webcite*

44. Nguyen CT, Roy G, Gauthier C, Galanis N: Heat transfer enhancement using Al$_2$O$_3$-water nanofluid for an electronic liquid cooling system.*Appl Therm Eng* 2007, 27:1501-1506.

45. Tseng SC, Lin CW, Huang K: Heat transfer enhancement of nanofluids in rotary blade coupling of four wheel-drive vehicles.*Acta Mechanica* 2005, 179:11-23.

46. Kulkarni DP, Vajjha RS, Das DK, Oliva D: Application of

aluminum oxide nanofluids in diesel electric generator as jacket water coolant.*Appl Therm Eng* 2008, 28:1774-1781.

47. Wu S, Zhu D, Li X, Li H, Le J: Thermal energy storage behavior of Al_2O_3-H_2O nanofluids.*Thermochim Acta* 2009, 483:73-77.

48. Wang XQ, Mujumdar AS: A review on nanofluids - part II: experiments and applications.*Braz J Chem Eng* 2008, 25:631-648.

Chapter 4

Rheological Profile Across the NE Japan Interplate Megathrust in the Source Region of the 2011 Mw9.0 Tohoku-oki Earthquake

Ichiko Shimizu

Department of Earth and Planetary Science, University of Tokyo, Bunkyo-ku, Tokyo 113-0033, Japan

ABSTRACT

A strength profile across the NE Japan interplate megathrust was constructed in the source region of the 2011 Tohoku-oki earthquake (M_w 9.0) using friction, fracturing, and ductile flow data of the oceanic crustal materials obtained from laboratory experiments. The depth-dependent changes in pressure, temperature, and pore fluid pressure were incorporated into a model. The large tsunamigenic slips during the M9 event can be explained by a large gradient in fault strength on the up-dip side of the M9 hypocenter, which was located 17 to 18 km beneath sea level. A large stress drop (approximately 80 MPa) induced by the collapse of a subducted seamount possibly triggered the M9 earthquake. In the deep (>35 km) part of the thrust fault, where M7-class Miyagi-oki earthquakes have repeatedly occurred, plastic deformation occurs in siliceous rocks but not in gabbroic rocks. Thus, the asperity associated with the M7-class earthquakes was most likely a gabbroic body, such as a broken seamount, surrounded by siliceous sedimentary rocks. The conditionally stable nature of the surrounding region can be explained by the frictional behavior of wet quartz in the brittle-ductile transition zone. In contrast to the deep M7-class asperity, the M9 asperity (i.e., a region that was strongly coupled before the M9 Tohoku-oki earthquake) extended to a large area of the plate interface because shear strength is relatively insensitive to litho-logical variation at intermediate depths. However, the along-arc extension of the M9 asperity was constrained by fluid-rich regions on the plate interface.

BACKGROUND

Before the 2011 Tohoku-oki earthquake (M_w 9.0), large earthquakes off the Pacific coast of NE Japan were thought to occur at specific areas, named asperities, on the subducting plate interface (Yamanaka and Kikuchi 2004; Matsuzawa et al. 2004). The term 'asperity' originally refers to a real contact on a sliding surface that is strong enough to sustain frictional stress. By analogy, seismic asperities are defined as patchy areas on a fault surface that prevent slippage during interseismic periods but cause a large slip upon failure (Scholz 2002; Okada et al. 2005). From the viewpoint of fault mechanics, asperities are interpreted as unstable (velocity-weakening) regions that are associated with repeated earthquake nucleation (Pacheco et al. 1993; Matsuzawa et al.2004).

In the NE Japan subduction zone, a typical asperity was located offshore of Miyagi (the Miyagi-oki area), where M7-class earthquakes occurred in 1936, 1978, and 2005, with a recurrence interval of approximately 37 years (Kanamori et al. 2006; Figure 1).

As the asperity was not completely broken during the 2005 event (Okada et al. 2005), another Miyagi-oki earthquake was predicted to occur in the near future (Iinuma et al. 2011). However, the M9 earthquake occurred on the up-dip side of the rupture areas of the previous M7-class Miyagi-oki earthquakes, where no asperities had been previously identified during seismic and geodetic observations (Yamanaka and Kikuchi 2004; Hashimoto et al. 2009). The slip area initially propagated trenchward and then to the deeper zones of the thrust fault (Koketsu et al. 2011; Honda et al. 2011; Yagi and Fukahata 2011; Yokota et al.2011). The area outside of the asperity associated with the earlier M7-class earthquakes was formerly assigned to a region of aseismic slip (Matsuzawa et al. 2004), but this region, as well as the asperity itself, was ruptured during the M9 event (Iinuma et al. 2011).

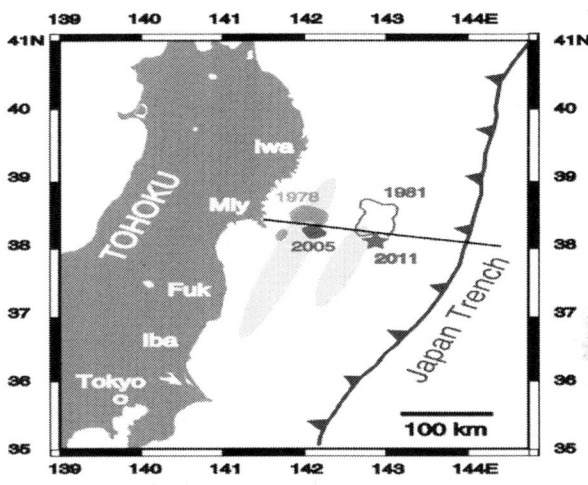

Figure 1: Index map of the Tohoku-oki subduction zone, NE Japan.The black line shows the location of the MY102 seismic survey line (Ito et al. 2005; Miura et al. 2005). The red star shows the epicenter of the great 2011 Tohoku-oki earthquake (M_w9.0) on March 11, 2011, which was determined by JMA, and green area indicates the large coseismic slip areas of the 1978 (M_w7.4) and 2005 (M_w7.2) Miyagi-oki earthquakes (Yamanaka and Kikuchi 2004; Okada et al. 2005). The coseismic slip area of the 1981 (M_w7.0) earthquake (Yamanaka and Kikuchi 2004) is outlined by a

blue line. The low-V and low- V_p/V_s domains observed beneath the plate interface (Matsubara and Obara 2011) are indicated by orange ellipses. Iwa, Iwate; Miy, Miyagi; Fuk, Fukushima; Iba, Ibaraki Prefecture.

To account for the occurrence of the gigantic M9 earthquake on the interplate megathrust where M7-class events repeatedly occur, simulation models incorporating complicated interactions of asperities have been developed based on the rate- and state-dependent friction (RSF) law (Hori and Miyazaki 2011; Kato and Yoshida 2011; Shibazaki et al. 2011; Mitsui et al. 2012; Noda and Lapusta 2013). The conditionally stable nature of the thrust fault outside the M7-class asperities has been reproduced in these numerical models. However, the large slips during the M9 event have been explained differently within these various models because of a lack of knowledge regarding the frictional properties and state of stress at the plate interface. This study aims to constrain the stress states on the NE Japan interplate megathrust using a rock-rheology-based approach. Herein, the effects of lithology, temperature (T), and pressure (P) on the strength of the subduction plate boundary are examined and the differences between M7- and M9-class earthquake asperities are discussed.

The strength of the continental and oceanic lithospheres has been discussed previously using the frictional law for brittle deformation and the dislocation creep flow laws for ductile deformation (Kohlstedt et al. 1995; Scholz 2002). This method was applied to the NE Japan subduction zone by Shimamoto (1989, 1993). However, the plate interface down to a depth of 30 km, including the hypocenter of the gigantic 2011 earthquake, was assigned to an a priori aseismic zone, as no distinct seismic activity had been previously identified in this region. The present study reconstructs the strength profile across the Tohoku-oki megathrust with special attention to the rheological properties of the oceanic crust materials, which include siliceous and clay-rich sediments, and basic volcanic and plutonic rocks, which are sandwiched between the overriding plate and the subducting slab. In subduction zones, H_2O-rich fluids are continuously supplied from the trench axis as pore water and from dehydrating zones in the subducting slab. Hence, the influence of pore fluid pressure and the chemical effects of water were incorporated into the present strength model.

METHODS

Geophysical Structures

Miura et al (2005) and Ito et al. (2005) investigated the velocity structure along a seismic survey line that transects the source region of the M9 Tohoku-oki earthquake (Figure 1). Figure 2 shows a simplified structural profile and interpretation of the forearc region. The upper crust consists of accreted sediments and sedimentary rocks including a Cretaceous-Tertiary sequence (Miura et al.2005; Tsuji et al. 2011). The hypocenter of the M9 earthquake was located at the base of the lower crust. Projection of the epicenter (38.103°N, 142.861°E, which was determined by the Japan Meteorological Agency, JMA) onto the plate interface yields a focal depth of 17 to 18 km below sea level. The epicenter relocated by Zhao et al (2011) is 38.107°N, 142.916°E, which is very close to that determined by JMA. The M7-class Miyagi-oki earthquakes occurred at the base of the mantle wedge. The oceanic Moho is traceable to the subducting slab (Miura et al. 2005).

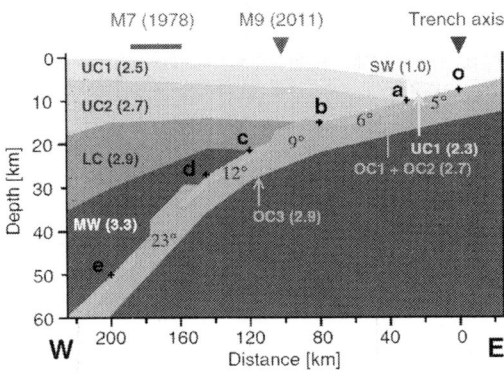

Figure 2: Simplified structural model of the forearc region offshore of Miyagi prefecture. The model is based on the density structure and the

seismic reflection images obtained along the black line shown in Figure 1 (Miura et al. 2005; Ito et al. 2005). Numbers in parentheses indicate the densities in gram per cubic centimeter. UC, upper crust; LC, lower crust; MW, mantle wedge; OC, oceanic crust; SW, seawater. Numbers on the subducting plate denote dip angles between reference points (o to e). The location of the asperity of the 1978 Miyagi-oki earthquake, the epicenter of the 2011 Tohoku-oki earthquake (M9), and the trench axis are shown above the cross section. The 'bending' structures in segments d-e and b-c (Ito et al. 2005) are interpreted as broken and unbroken seamounts, respectively (see text for details).

The seismic reflection images revealed 'bending' of the oceanic plate at the same depth as the M9 hypocenter and at the eastern edge of the asperity of the Miyagi-oki earthquake (Ito et al. 2005). These bending structures are most probably subducted seamounts, as gravity data show a positive anomaly at the M9 hypocentral zone (Figure eleven (a) of Miura et al. 2005). Three-dimensional (3D) tomography of P- and S-wave velocities (V_p and V_s, respectively) identified low-velocity (V) and low- V_p/V_s anomaly zones beneath the plate interface (Matsubara and Obara 2011; Figure 1). Kennett et al (2011) and Tajima and Kennett (2012) also detected a V_s anomaly on the trenchward side. The presence of pore fluids and serpentinite (e.g., Watanabe 2007; Peacock et al.2011) is a possible cause of the low-V anomaly, but the low- V_p/V_s anomaly is more likely explained by partial melting and/or a heat source associated with volcanic activity at seamounts. The convex configuration of the seafloor geography at the southern part of the trenchward anomaly zone offshore of Fukushima also suggests that this zone has been the site of seamount subduction (Matsubara and Obara 2011). The landward velocity anomaly zone is located slightly to the west of the deep bending structure identified by Ito et al. (2005). It is possible that this deep bending is a faulted seamount that is stuck at the base of the mantle wedge and that the low-V zone represents a vent, as shown schematically in Figure 2.

The NE Japan subduction zone is characterized by rapid subduction of the old (approximately 130 Ma) Pacific plate (Wada and Wang 2009). The thermal gradient along the plate interface is comparable to that of a high-P/low-T metamorphic path, and the deep parts of the fault zone beneath the island-arc Moho correspond to lawsonite-blueschist facies metamorphism (Hacker et al. 2003; Omori et al. 2009). Field studies in high-P/low-T metamorphic belts suggest that basaltic lavas and pyroclastic rocks that constitute oceanic layer 2 are largely replaced by metamorphic minerals such as glaucophane, lawsonite, and chlorite, whereas coarse-grained pyroxene is well preserved in gabbros and dolerites that constitute oceanic layer 3 and subducted seamounts (e.g., Agata 1994). The total thickness of the subducting oceanic crust is about 7 km and the lower 5 km is oceanic layer 3 (Miura et al. 2005). In the rheological model described below, oceanic layer 3 and broken seamounts are represented by gabbroic layers with thicknesses of 5 and 7 km, respectively.

Frictional and Fracture Strengths

The shear strength (τ) of rocks under dry or water-saturated conditions that contain pre-existing fault planes is generally expressed as

$$\tau = \tau_0 + \mu(\sigma_n - \alpha P_p)$$

(1)

where τ_0 is the cohesive strength, σ_n is the normal stress on the fault surface, μ is a constant, P_p is the pore pressure, and α is a factor between 0 and 1 (Paterson and Wong 2005). In the case of perfectly brittle deformation, $\alpha = 1$ can be applied and the above equation reduces to

$$\tau = \tau_0 + \mu\sigma_n'$$

(2)

where $\sigma_n'(=\sigma_n - P_p)$ is the effective normal stress.

The steady-state friction coefficient (τ/σ_n') of quartz gouge under water-saturated conditions ranges from 0.65 to 0.75 at 25°C to 200°C (Chester and Higgs 1992; Chester 1995; Nakatani and Scholz 2004a) and that of wet granite gouges is about 0.7 at temperatures up to 300°C and $\sigma_n' = 400$ MPa (Blanpied et al. 1991,

1995). Hydrothermal experiments of gabbro gouge conducted by He et al. (2007) yield τ/σ'_n=0.70 to 0.75 at temperatures up to 581°C (Figure 3a). These values are slightly smaller than $\mu = 0.85$ of dry rocks (Byerlee 1978). In this study,

$$\tau_0 = 0 \text{ MPa} \quad \text{and} \quad \mu = 0.7 \quad \text{at} \quad \sigma'_n < 500 \text{ MPa}$$

(3)

And

$$\tau_0 = 50 \text{ MPa} \quad \text{and} \quad \mu = 0.6 \quad \text{at} \quad \sigma'_n > 500 \text{ MPa}$$

(4)

are used for both siliceous and basic rocks. The frictional law defined by Equations 2 to 4 is shown as a dotted blue line in Figure 3a.

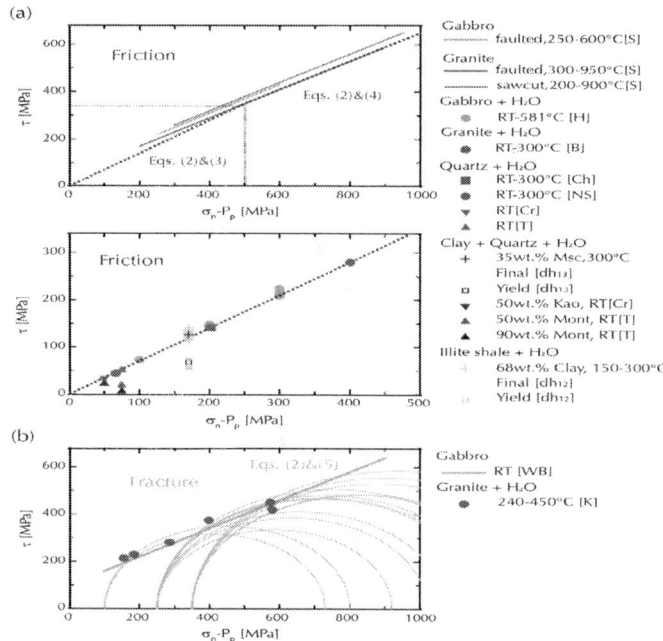

Figure 3: Brittle strength of rocks and gouges at various pressures and temperatures. (a) Frictional strength of dry rocks (upper) and gouges (low-

er) plotted in the dotted rectangular area of the upper diagram. The dotted blue line represents the frictional strength model given by Equations 3 and 4. (b) Mohr circles and fracture strength of intact rocks. The thick green line is the fracture strength model given by Equation 5. RT, room temperature; S, Stesky et al. (1974); H, He et al. (2007); B, Blanpied et al. (1995); Ch, Chester (1995); NS, Nakatani and Scholz (2004a); T, quartz+montmorillonite (Mont) mixtures at a shear displacement of 2.2 mm after Takahashi et al. (2007). The clay contents in weight percentage were given by Takahashi (personal communication). Cr, quartz+kaolinite (Kao) mixtures at 5% shear strain after Crawford et al. (2008); dH13, quartz+muscovite (Msc) mixture (den Hartog et al. 2013); dH12, illite-rich shales (den Hartog et al. 2012); WB, Wong and Biegel (1985); K, Kato et al. (2003).

Laboratory experiments indicate that clay and serpentine minerals have friction coefficients smaller than a μ of approximately 0.7 for ordinary minerals (Byerlee 1978; Moore et al. 1997; Takahashi et al. 2011); swelling clays (smectite) and a low-T serpentine polymorph of chrysotile have especially low friction coefficients (0.1 to 0.2). The friction coefficients of clay-rich gouges depend on the total amount of clay and the clay mineralogy. Figure 3a plots the strength of mixed gouges of quartz and montmorillonite, a swelling clay of the smectite group, and of quartz and kaolinite at room temperature under water-saturated conditions (Takahashi et al. 2007; Crawford et al. 2008). Smectite-rich clays have extremely low friction coefficients (0.1 to 0.2). At deep parts of accretionary prisms, however, smectite becomes unstable and transforms to illite at temperatures above 100°C to 150°C and then to muscovite at higher temperatures (> 300°C). High-pressure hydrothermal experiments using a rotary shear apparatus suggested that the friction coefficients of illite- and muscovite-rich gouges are considerably larger than those of clay-rich gouges under low- σn conditions (den Hartog et al. 2012, 2013). In their experiments, however, μ of quartz-phyllosilicate gouges increased with shear displacement and the steady-state values of μ were not attained in most runs, although the final displacements were very large (25 to 74 mm). Progressive crushing of quartz grains and preferential squeezing out of phyllosilicates are the possible causes of this slip-hardening behavior (den Hartog and Spiers 2013). The effects of serpentine minerals were neglected in the present study because chrysotile is metastable and not significant under the P-T conditions of the mantle wedge (Evans 2004). In addition, the velocity structures obtained from the region offshore of Miyagi suggest that the mantle wedge in this area is not intensively serpentinized (Miura et al. 2005; Yamamoto et al. 2008).

Shear failure of intact rocks can also be described by Equation 2. In this case, Equation 2 represents the Mohr-Coulomb criterion under water-saturated conditions, and μ is an internal friction. The failure criterion of intact gabbro (Wong and Biegel 1985) at room dry conditions (shown by the dark green line in Figure 3b) is approximated by Equation 2 and

$$\tau_0 = 100 \text{ MPa} \text{ and } \mu = 0.6$$

(5)

The failure conditions of intact granite under high-pressure hydrothermal conditions (Kato et al.2003) are well approximated by Equations 2 and 5 (shown as a thick green line in Figure 3b). In the absence of experimental data for gabbro under the water-saturated conditions, these parameters and Equation 2 were used to describe the fracture strength of gabbro in subduction zones.

Plastic Strength

Dislocation Creep

High-temperature dislocation creep of rocks and minerals is generally described by a power law relationship:

$$\dot{\varepsilon} = A(\sigma_1 - \sigma_3)^n \exp\left(-\frac{Q}{RT}\right)$$

(6)

where $\dot{\varepsilon}$ is the strain rate, σ_1 and σ_3 are the maximum and minimum principal stresses, respectively, Q is the activation energy, R is the gas constant, and A and n are material constants. At very high stresses and relatively low temperatures, the power law relationship breaks down and the flow law is instead approximated by an exponential law (Frost and Ashby 1982; Kohlstedt et al.1995). However, the exact form of the exponential law has not been established for crustal materials such as quartz. The power-law breakdown was therefore not considered in this section, but its influence was indirectly incorporated into the model of the brittle-ductile transition zone described below.

Intracrystalline slip is nearly independent of the intermediate principal stress (σ_2). Thus, flow stress during plastic deformation can be described roughly by Tresca's criterion:

$$\tau = \frac{1}{2}(\sigma_1 - \sigma_3)$$

(7)

Equations 6 and 7 yield shear strength against plastic flow:

$$\tau = \frac{1}{2}\left(\frac{A}{\dot{\varepsilon}}\right)^{\frac{1}{n}} \exp\left(\frac{Q}{nRT}\right)$$

(8)

In studies of lithospheric strengths, rheological properties of mantle peridotites are generally represented by the frictional and flow strengths of olivine (Kohlstedt et al. 1995). However, if the oceanic crust materials at the top of the subducting slab are weaker than olivine, the movements of the interplate thrust faults are governed by these weak materials. The following sections evaluate the flow strengths of the siliceous sedimentary and basic rocks that constitute the oceanic crust. In the interplate seismogenic zone (<60 km in depth) (Pacheco et al. 1993), both the mantle wedge and the slab mantle can be treated as rigid bodies because the dislocation creep of olivine is not activated at temperatures below 600°C at the relevant geological strain rates (Shimamoto1993; Kohlstedt et al. 1995).

Wet Quartz

Plastic deformation of the siliceous sedimentary layer that constitutes the uppermost part of the subducting slab was approximated by dislocation creep of wet quartz. Figure 4 shows the flow stress of wet quartz at a constant strain rate ($\dot{\varepsilon}$ =10−12s−1), which was estimated from flow laws determined by gas-medium apparatuses at a 300 MPa confining pressure (Paterson and Luan1990; Luan and Paterson 1992; Rutter and Brodie 2004; Mainprice and Paterson 2005) and by a Griggs apparatus with molten-salt cells at higher P_c (approximately 1.5 GPa) (Gleason and Tullis1995). Large variations in flow stress arise from differences in P_c, the starting materials used, and the chemical environments.

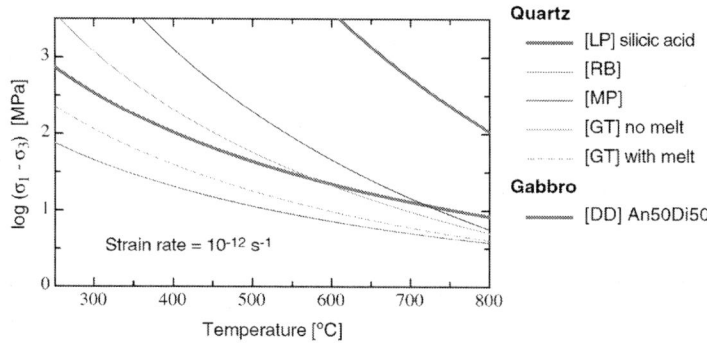

Figure 4: Plastic strength of wet quartz and gabbro. Extrapolation of experimentally determined dislocation flow laws of wet quartz and gabbro to a natural strain rate (˙ =10−12s−1) are shown. LP, fine quartz aggregate synthesized from silicic acid (Paterson and Luan 1990; Luan and Paterson 1992); MP, Mainprice and Paterson (2005); RB, fine quartz aggregate synthesized from Brazilian quartz (Rutter and Brodie 2004); GT, Black Hills quartzite (Gleason and Tullis 1995); DD, synthetic gabbro (An50Di50, 'wet') of Dimanov and Dresen (2005).

To constrain Q and the water fugacity factor in the flow law of wet quartz, Hirth et al. (2001) investigated the deformation conditions of quartz mylonites and extrapolated laboratory data of quartzite flow laws to the natural deformation conditions. The semi-empirical flow law of Hirth et al. (2001) was, however, not used in the present model because the tectonic setting, grain size, and water contents of the quartzite samples are far different from those of chert in subduction zones. Moreover, they used a temperature-independent grain size piezometer to estimate differential stress during natural deformation, although considerably large effects of temperature are theoretically predicted (De Bresser et al. 2001; Shimizu 2008, 2012).

The pressure-dependence of dislocation creep is generally expressed in terms of water fugacity fH2O and activation volume Ω. The exact value of Ω has not been determined for quartz, but it is expected to be on the same order as the molar volume of quartz ($v_{qtz} = 22.7$ cm^3 mol^{-1}). Taking $\Omega = v_{qtz}$, for example, then the increase in Q coincident with a pressure increase (P) of 1 GPa caused by the activation volume term (Ω P) is 20 to 30 kJ/mol^1, which is far smaller than the uncertainty in the experimental determination of Q for wet quartz. Thus, the direct effect of pressure imposed by the activation volume term was neglected in the rheological model discussed here.

The effect of water fugacity was introduced to the pre-exponential constant in Equation 6 as A$\propto$fmH2O, where the exponent m is dependent on the water-related defect species within the crystals. Assuming equilibrium relationships between dissolved 'water' in the crystal and the surrounding vapor phase, Paterson (1989) derived m values of 1 to 2 for quartz. High-PTexperiments with quartz suggested that m $\leq$ 1 (Chernak et al. 2009). Under experimental conditions at temperatures higher than 900°C, water is in a vapor phase and fH2O is nearly equal to water pressure. Accordingly, the flow stress of wet quartz decreases with increasing pressure. In high-P/low-T metamorphic conditions along the subducting plate interface, however, H_2O fluids exist as liquid water with densities of around 1 g/cm^3 (Burnham et al. 1969). Thus, the water fugacity correction is not applicable to subduction zones. Ito and Nakashima (2002) reported that the water content of chert gradually decreases with increasing metamorphic grade, although fH_2O increases with increasing temperature and pressure. This means that equilibrium concentrations of water-related species are not attained under low-grade metamorphic conditions. Ito and Nakashima (2002) also showed that most of the water in chert is distributed along grain boundaries. Therefore, direct use of the flow laws of chert, or of synthetic quartz aggregates that are similar in water contents and grain size to chert, would be most suitable for application to subduction zones. The strength of flint determined by Mainprice and Paterson (2005) is considerably lower than the strength of fine-grained synthetic quartz aggregates (Paterson and Luan 1990; Luan and Paterson1992; Rutter and Brodie 2004). The weakness of the flint samples may be related to the formation of localized shear bands that were observed in the deformed samples. The total water contents of the synthetic quartz aggregates used by Rutter and Brodie (2004) were relatively small (e.g., 100 to 200 $H/10^6$ Si after deformation), and the pore fluids were not saturated with H_2O during deformation (see discussion by Shimizu (2008)). A series of experiments conducted by Paterson and Luan (1990) and Luan and Paterson (1992) used starting materials that were synthesized from powders of natural quartz, silicic acid, and silica gel; silicic-acid- and silica-gel-origin samples have water contents (approximately 10^4 $H/10^6$ Si) comparable to those of weakly metamorphosed cherts (Ito and Nakashima 2002).

Luan and Paterson (1992) identified possible impurity hardness effects in experiments using silica-gel-origin samples. As such, the flow law of silicic-acid-origin samples without correcting for Ω and fH2O was used in this study. Paterson and Luan (1990) presented the flow law parameters of quartz in a modified form of Equation 6:

$$\frac{\dot{\varepsilon}}{\dot{\varepsilon}'} = \left(\frac{\sigma_1 - \sigma_3}{\sigma_1' - \sigma_3'}\right)^n \exp\left[-\frac{Q}{R}\left(\frac{1}{T} - \frac{1}{T'}\right)\right]$$

(9)

where differential stress at a reference state ($\dot{\varepsilon}'$=1×10−5s^{-1} and T'= 1,300 K) was given as $\sigma_1' - \sigma_3'$=222 MPa for silicic-acid-origin samples. Revised values of Q and n given by Luan and Paterson (1992) are 152 kJ/mol[1] and n = 4.0, respectively.

Oceanic Crust

The flow stress of gabbro appropriate for oceanic layer 3 was calculated in Figure 4 using the flow law for a synthetic aggregate of anorthite and diopsite presented by Dimanov and Dresen (2005). The effective viscosity ($=\tau/\varepsilon$) of synthetic gabbro is more than two orders of magnitude higher than that of wet quartz at temperatures less than 600°C. Plastic deformation of oceanic layer 3 was therefore ignored in the strength model.

In subduction zones, basaltic lava and pyroclastic rocks are recrystallized into metamorphic minerals that are characterized by blueschists or greenschists. Currently, little is known about the flow strength of basic schists. In general, quartz is considered to be the weakest of the major constituent minerals in metamorphic rocks. Nevertheless, field observations of low-grade metamorphic rocks suggest that the strain magnitudes of metabasites are about the same as those of chert and shale in the same area (Shimizu 1988; Shimizu and Yoshida 2004). Here, the shear deformation of oceanic layers 1 and 2 is described in terms of quartz rheology and an effective thickness w (Figure 5); $w = 2$ km means that the composite of layers 1 and 2 with a total thickness of 2 km obeys the flow law of wet quartz. In contrast, $w = 0.3$ km indicates that the uppermost quartz-rich layer with a thickness of 0.3 km deforms identically to wet quartz, whereas the lower basaltic layer behaves as a rigid body. Using the plate convergence rate $V_0 = 83$ mm/year ($= 2.6 \times 10^{-9}$ m/s) at the Japan Trench (Wada and Wang 2009) and assuming $w = 1$ km, for example, the strain rate of the plastic part of the oceanic crust was calculated as $\cdot = V_0/w = 2.6 \times 10^{-12} s^{-1}$.

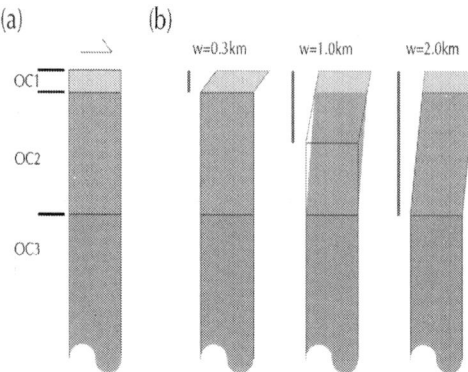

Figure 5: Deformation partitioning models for the subducting oceanic crust. (a) Initial state. (b) Shear deformation of the oceanic sedimentary (OC1) and basaltic (OC2) layers at different effective thicknesses w (shown as blue bars). No deformation is considered to occur in the gabbroic layer (OC3).

Rheological Model

Plate Boundary Model

Subduction zone megathrusts generally have curved interfaces that are not always favorably oriented to satisfy the yield criterion shown in Equation 1. Herein, it is assumed that σ_1 is horizontal and perpendicular to the trench axis and σ_3 is vertical. Then, the normal and shear stresses on the plate boundary thrust fault with a dip angle θ are written as

$$\sigma_n = \frac{\sigma_H + \sigma_V}{2} - \frac{\sigma_H - \sigma_V}{2} \cos 2\theta$$

$$\text{(10)}$$

$$\tau = \frac{\sigma_H - \sigma_V}{2} \sin 2\theta$$

$$\text{(11)}$$

where σ_V and σ_H are the vertical and horizontal rock stresses, respectively. Combining Equations 1, 10, and 11, the frictional or fracture strength of the thrust fault can be obtained as a function of ϑ:

$$\tau = \tau_0^* + \mu^*(\sigma_V - \alpha P_p) \tag{12}$$

Where

$$\tau_0^* = \frac{\tau_0 \sin 2\theta}{\sin 2\theta - (1 - \cos 2\theta)\mu} \tag{13}$$

And

$$\mu^* = \frac{\mu \sin 2\theta}{\sin 2\theta - (1 - \cos 2\theta)\mu} \tag{14}$$

The parameters given in Equations 3 and 4 were used for the friction of quartz-rich sedimentary rocks and gabbros, whereas Equation 5 was applied for the fracture strength of gabbro.

Using a simplified density model of the forearc region (Figure 2) and the structural parameters listed in Table 1, the distribution of σ_V on the plate interface was calculated (shown as a lithostatic line in Figure 6a). The density of H_2O was taken to be 1.0 g/cm^3 everywhere within the model. Assuming $\alpha = 1$, the fracture and frictional strengths of siliceous and gabbroic rocks under a hydrostatic pore pressure gradient (shown as a light blue line in Figure 6) were calculated, and these values are shown in Figure 7.

Table 1: Structural parameters for the thrust model

	Distance	Depth	Thickness (km)					ϑ_a
	(km)	(km)	SW	UC1	UC2	LC	MW	(deg)
o	0	7.5	7.5	0.0	0.0	0.0	0.0	-
a	30	10.0	5.0	5.0	0.0	0.0	0.0	5.5
b	80	15.0	2.0	5.0	8.0	0.0	0.0	6.7
c	120	21.5	1.5	5.0	8.0	7.0	0.0	11.1
d	145	27.0	1.0	5.0	8.0	7.0	6.0	16.7
e	200	50.0	0.0	5.0	10.0	15.0	20.0	25.3

[a]To determine θ, the positions of the reference points a to e were fitted by a sixth polynomial function. Symbols are explained in Figure 2.

Shimizu Earth, Planets and Space 2014 66:73 doi:10.1186/1880-5981-66-73.

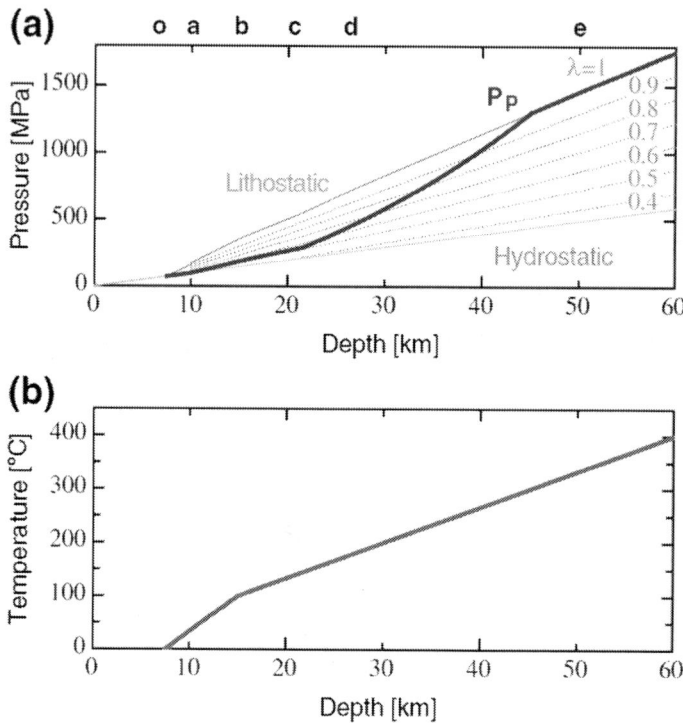

Figure 6: Pressure (a) and temperature (b) models for the Tohoku-oki megathrust. The green line represents the lithostatic pressure σ_v, and the light blue line shows the hydrostatic pressure gradient. The pore pressure ratio λ is indicated by dotted lines. The thick blue line represents the pore pressure model. P_p at the depth $z = 21.5$ to 45 km is given by $P_p = \sigma_v((1 - \beta)\lambda_c + \beta)$ and $\beta = (z - z_{up})/(z_{down} - z_{up})$, where $\lambda_c = 0.53$, $z_{down} = 45$ km, and $z_{up} = 21.5$ km.

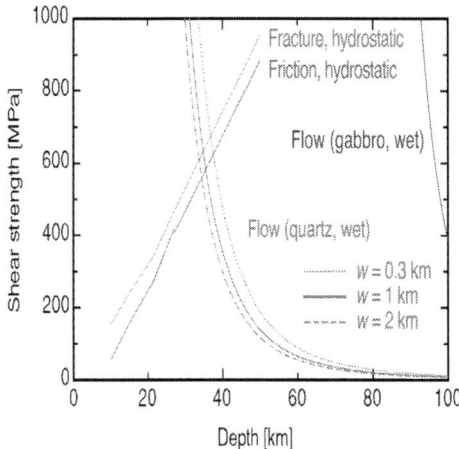

Figure 7: A 'simple' rheological model of the Tohoku-oki megathrust assuming a hydrostatic pore pressure gradient. The green and orange lines represent fracture and frictional strengths, respectively. The flow strengths of wet quartz and gabbro are shown by pink and purple lines, respectively.

The thermal models of the NE Japan subduction zone proposed by Omori et al. (2009) and Wada and Wang (2009) yield temperatures for the plate interface of about 200°C at a depth of 30 km and 400°C at a depth of 60 km, whereas Hacker et al. (2003) and Iwamori (2007) proposed higher-P/lower-T paths. In this study, a constant thermal gradient (= 100°C per 15 km) at depths greater than 15 km (Figure 6b) was assumed based on the results of Omori et al. (2009) and Wada and Wang (2009). This thermal model gives conservative estimates for plastic strengths of siliceous sedimentary rocks and subducted seamounts (shown as pink and purple lines in Figure 7, respectively).

Brittle-ductile Transition Zone

A conventional way to construct a strength profile along a fault zone is to connect frictional strength and flow stress lines at a crossover point. However, this two-mechanism envelope yields a steep peak that has not been observed in laboratory studies (Shimamoto 1986).

It has been discussed that the strength in the brittle-ductile transition zone gradually changes because of the onset of several different mechanisms such as semi-brittle flow, power-law breakdown, and pressure solution precipitation (Shimamoto 1993; Hirth and Tullis 1994; Chester 1995; Kohlstedt et al. 1995; Bos and Spiers 2002). Whatever the deformation mechanisms at microscopic scales are, the macroscopic changes from frictional sliding to plastic flow can be expressed using two parameters: τ_0 and μ. The frictional strength at shallow parts is represented by Coulomb's criterion with $\tau_0 = 0$ and $\mu > 0$ (Figure 8a). Increasing pressure and temperature cause an increase in τ_0 and a decrease in μ. In fully plastic deformation, μ reduces to zero (i.e., $\tau = \tau_0$) as represented by Tresca's criterion in Equation 8 (Figure 8b). Accordingly, τ^*_0 increases and μ^* decreases with increasing depth in the brittle-ductile transition zone.

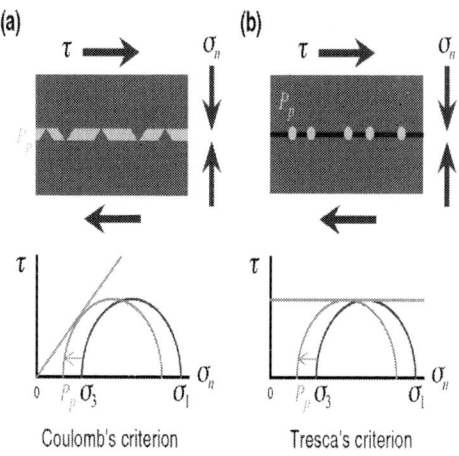

Figure 8: Schematic illustrations of fault interfaces (upper) and the influence of pore fluid pressure on yield criteria (lower). (a)Frictional slip governed by Coulomb's criterion. (b) Ductile deformation governed by Tresca's criterion. See text for details.

Experimental and theoretical research suggests that dissolution-precipitation of quartz becomes significant at temperatures above 150°C (Shimizu 1995; Chester 1995; Nakatani and Scholz2004b), and this corresponds to the base of the island-arc crust (point c) in the thermal model (Figure 6b). Hence, in the improved rheological profile shown in Figure 9, the down-dip side of point c was assigned to the brittle-ductile transition zone. A strain analysis of chert and mudstone that have undergone lower greenschist facies metamorphism (approximately 300°C) showed the dominance of intracrystalline plastic deformation (presumably dislocation creep) over pressure solution (Shimizu 1988). Microstructural observations in ductile shear zones suggest that quartz is predominantly deformed by dislocation creep at temperatures above 300°C (e.g., Stipp et al.2002). As a first approximation, the plate interface at depths greater than 45 km (approximately 300°C) was assigned as the perfectly plastic deformation zone of quartz, and $*_0$ and μ^* in the brittle-ductile transition zone were varied linearly as shown in Figure 9a. Figure 7 indicates that the temperature range of the brittle-ductile transition of gabbro is much higher than that of quartz. The modeling presented here therefore tentatively estimates that the brittle-ductile transition depth was 27 to 95 km; the upper limit corresponds to point b in Figure 2.

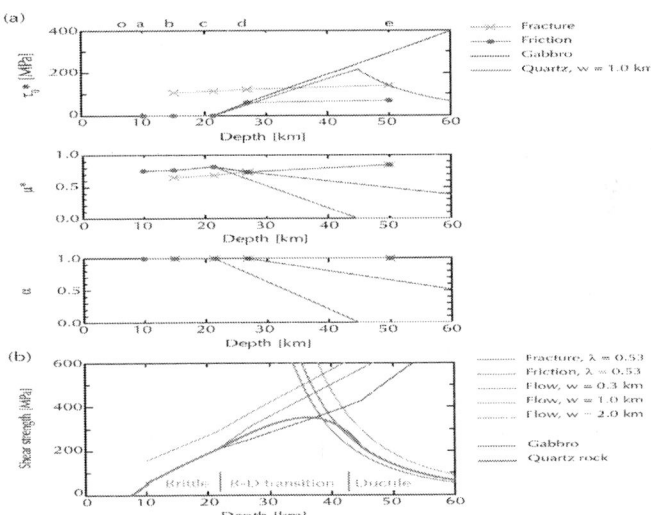

Figure 9: Improved rheological model of the Tohoku-oki megath-rust. Here the 'simple' rheological profile in Figure 7 is modified

using the pore pressure model shown in Figure 6 and considering the brittle-ductile transitional behaviors of quartz and gabbro. (a) Depth-dependent changes in model parameters τ^*_0, μ^*, and λ. Brittle-ductile transitional behaviors of gabbro and quartz are taken into account in purple and thick pink lines, respectively. (b) Shear strength along the interplate megathrust. The fracture and frictional strengths (shown as green and orange lines, respectively) of gabbro and siliceous rocks were calculated assuming $\lambda = 0.53$. The strength envelopes of quartz and gabbroic rocks are shown as thick pink and purple lines, respectively.

In the fully brittle regime, stress on the fault surface is supported by small solid-solid contact areas, i.e., 'asperities' in the original sense (Dieterich and Kilgore 1994). In this configuration, frictional strength is reduced with increasing pore pressure as expressed by Equation 1 with $\lambda = 1$ (Figure 8a). At high temperatures and pressures, solids at the real contact areas yield and pore spaces are reduced by dissolution-precipitation processes. This causes an increase in contact areas, which results in a decrease in a and τ_0. The fully plastic deformation regime was associated with tube- or inclusion-like pore geometries (Figure 8b). In this configuration, direct mechanical effects of pore pressure almost disappeared and an   of approximately 0 was attained. In the modeling presented here,   was changed linearly across the brittle-ductile transition zone (Figure 9a).

Pore Fluid Pressure

In the previous calculation (Figure 7), a simple case of the hydrostatic pore pressure gradient was applied for the frictional and fracture strengths. This model gives the upper limit for the strength profiles in the brittle zone. In actual accretionary zones, pore pressure and pore pressure ratios $\equiv P_p / _V$ are influenced by sedimentation and compaction rates and the presence of impermeable materials such as clay-rich sediments; hence, they generally deviate from hydrostatic trends. In the improved model described below, the pore pressure ratio at the base of the accretionary wedge estimated by the Coulomb wedge model (Davis et al. 1983) was used for segment b-c of Figure 6a. Davis et al. (1983) defined generalized pore pressure ratios $\lambda\tilde{}\equiv P\tilde{}_p/\sigma\tilde{}_V$ at the basal thrust, where $P\tilde{}p$ and $\sigma\tilde{}V$ are fluid and lithostatic pressures calculated from the seafloor, respectively. Applying $\mu = 0.85$ (Byerlee 1978) to basal friction (μ_b) and $\mu = 1.03$ to internal friction (μ_i), they derived $\lambda\tilde{}=0.5$ for the NE Japan subduction zone. The given parameters of μ_b and μ_i are greater than the range of friction coefficients determined in laboratories; however, using friction coefficients of wet gouges and dry rocks for basal and internal friction, respectively (i.e.,$\mu_b = 0.7$ and $\mu_i = 0.85$), almost the same results can be obtained in the Coulomb wedge model (see Figure two of Saffer and Bekins 2002). Adopting $\lambda\tilde{}=0.5$ for reference point b (Table 1), a value of $\lambda = 0.53$ was obtained.

The change in pore structures within the brittle-ductile transitional zone shown in Figure 8 would be accompanied by a reduction in porosity and a decrease in permeability that in turn would influence pore pressure and effective normal stress on the fault plane (Yoshida and Kato 2011). The dissolution-precipitation of quartz in siliceous rocks is considered to be the most effective mechanism of porosity reduction at high temperatures (>150°C). Herein, P_p is expressed as a simple function of the depth (z) that gradually approaches the lithostatic level in the brittle-ductile transition zone of quartz (Figure 6a). The maximum strength of the thrust fault depends on the assumed pore pressure distribution, but the general features of the strength envelope are not significantly changed.

RESULTS AND DISCUSSION

Strength Envelope

The strength envelopes of the oceanic sedimentary layer were outlined for the case of w = 1 km (Figure 9b) using the parameters τ^*_0, μ^*, and α described above (Figure 9a) and the P_p distribution shown in Figure 6a. The strength of the gabbroic rocks was calculated using the same approach. The frictional strengths of clay-rich sediments in the shallowest zone (segment o-a) were evaluated in Figure 10a. The friction coefficients of quartz-montmorillonite (Takahashi et al.2007) and quartz-kaolinite gouges (Crawford et al. 2008) were used for the lower limit (denoted as smectite clay) and the upper limit (denoted as smectite-free clay), respectively. These values were also applied to segment a-b (shown as dotted lines in Figure 10a), although the pressure and temperature ranges exceeded the experimental conditions.

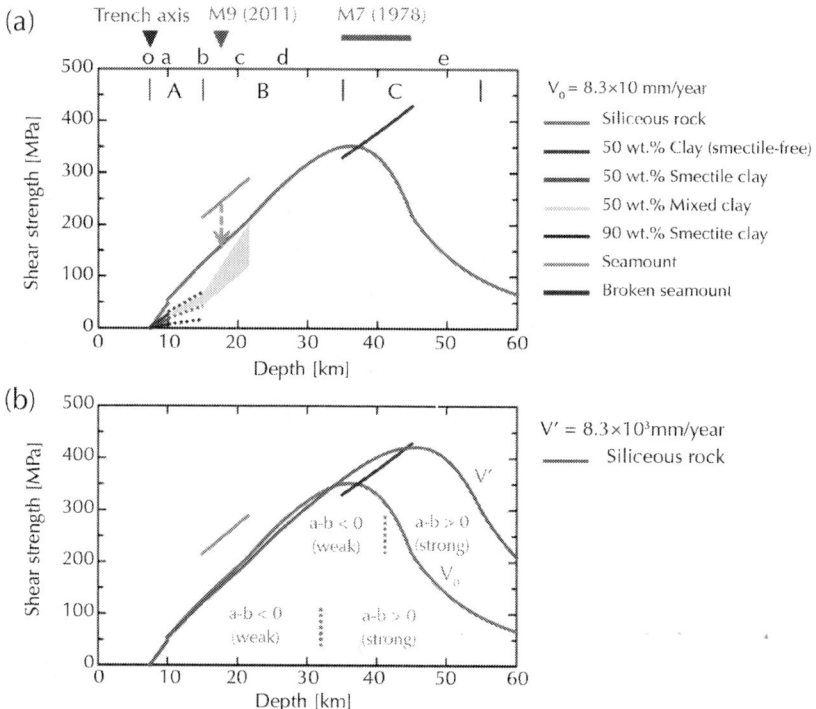

Figure 10: Rheological model of the M9 earthquake generating fault. (a) Strength envelope of the thrust fault during interseismic periods (V = 83 mm/year). The frictional and fracture strengths of gabbro (Figure 9b) are used for broken and unbroken seamounts. A green arrow indicates a stress drop induced by a collapse of a sea-

mount at the M9 hypocenter. A to C, rupture domains defined by Lay et al. (2012). (b)Velocity dependence of shear strength. The orange line shows the strength envelope of quartz at a faster slip rate (V' = V × 10²). Here $\tau = 0$ and $\mu = 0.67$ are used in the brittle zone, and a-b is defined with a change in steady-state friction coefficient (μ_{ss}) as a - b = $\Delta\mu_{ss}$/Δ lnV. The depth of the brittle-ductile transition zone is taken to be 21.5 to 55 km.

The smectite-rich sediment with a total clay content of approximately 90 wt.% is comparable to pelagic mud in the décollement zone that was obtained at a drilling site close to the trench axis (Ujiie et al. 2013). At the base of the accretionary wedge, the main thrust fault penetrates the horst-and-graben structure of the subducting oceanic plate (Tsuji et al. 2011). Hence, not only pelagic sediments but also trench materials with various clay contents are involved in the main thrust zone. The frictional strength of the thrust fault at the depth of point a varies in the range of 6 to 50 MPa. The weakest material of smectite gradually transforms to illite at elevated temperatures and it almost disappears within the temperature range of 100°C to 150°C (Kimura et al. 2012; Saffer and Tobin 2011), which corresponds to segment b-c on the plate interface (see Figure 6b). The range of frictional strength for 50 wt.% clay gouges that subducted to the depth of point c is represented by a light blue area in Figure 10a.

The nucleation site of the M9 earthquake would not have been largely covered with clay-rich sediments because both smectite- and illite-rich gouges exhibit velocity-strengthening behaviors over a wide range of pressure and temperature conditions (Saffer and Marone 2003; den Hartog et al. 2012; den Hartog and Spiers 2013). It is likely that the uppermost part of the oceanic crust is underplating at this depth and that the main thrust plane penetrates the chert-basalt sequence. Applying μ=0.7 for siliceous and basic rocks, the fault strength was estimated to be approximately 100 MPa at the base of the forearc upper crust (point b) and approximately 150 MPa at the hypocenter of the M9 earthquake. Because the value determined at point b was extrapolated into segment b-c in this model, the strength at the M9 hypocenter may be somewhat overestimated. The maximum value of τ on the interplate megathrust was not well constrained in the present model because of the uncertainty in rheological parameters in the brittle-ductile transition zone and pore pressure distributions. Future experimental studies and hydrologic modeling are needed to determine parameters τ_0, μ^*, α, and λ. In the following sections, we only use the general features of the brittle-ductile transition zone to discuss seismogenesis on the M9 earthquake-generating fault.

Asperities on Megathrust Faults and the Tohoku-oki M9 Earthquake

A well-established concept for seismic activity in the continental crusts is that large earthquakes nucleate at the deepest part of the brittle deformation zone (Sibson 1983; Scholz 1988, 2002). However, this concept has not been successfully applied to subduction zone megathrusts; for example, using the rheological data of olivine, the down-dip limit of the brittle zone has been estimated to be around 60 to 70 km (Shimamoto 1993), whereas the hypocenters of M9-class earthquakes are concentrated at shallower depths (<35 km depth) (Lay et al. 2012). The rheological model outlined here is based on the idea that the strength of the interplate megathrust is primarily controlled by the crustal materials that are weaker than mantle olivine. It is noteworthy that the hypocenter of the M9 Tohoku-oki earthquake was located at a deep part of the brittle zone of wet quartz (Figures 9b and 10a).

A joint inversion model of teleseismic, strong motion, and geodetic data (Koketsu et al. 2011) for the Tohoku-oki earthquake demonstrated that the coseismic slip was largest at the hypocentral area corresponding to segment b-c in Figure 2. This means that large elastic strain accumulated in the hypocentral area before the M9 earthquake. Global positioning system (GPS) data indicated that a wide area of the fault plane including the M9 hypocenter was tightly coupled before the 2011 earthquake (Nishimura et al. 2000; Suwa et al. 2006; Figure 11). These observations are well explained by the brittle and strong nature of the interplate thrust fault at intermediate depths (15 to 20 km). By contrast, tsunami inversions (e.g., Koketsu et al. 2011; Yokota et al. 2011) and a geodetic model including data from seafloor GPS stations (Iinuma et al. 2012) indicate that the largest slip area was close to the trench axis. Inversion of high-rate GPS data (Yue and Lay 2011) also indicates that the peak slip areas were located in both shallow and intermediate zones. This suggests that the shallow tsunamigenic zone was deformed passively and inelastically without affecting ground motions (Koketsu et al. 2011).

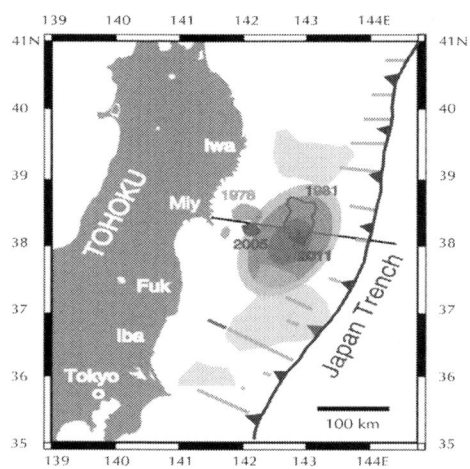

Figure 11: Relationship between interplate coupling and the velocity structure within the overriding plate. The area of strong interplate coupling (Nishimura et al. 2000) is colored pink. The light blue areas show the low-V and high- V_p/V_s domains above the interplate megathrust (Zhao et al. 2011). Brown lines show the distribution of the low-velocity sedimentary units observed by Tsuru et al. (2002). The blue line indicates the low-velocity part of the mantle wedge (Miura et al.2003). Other symbols are explained in Figure 1.

Geophysical observations suggest the existence of a subducted seamount in the hypocentral zone of the M9 earthquake, as described above. The bending structure of the subducting oceanic plate (Ito et al. 2005) has a width of approximately 50 km, which is the typical size of seamounts (Tsuru et al. 2002), and the epicenter of the M9 mainshock event is located on the western edge of the bending structure as illustrated in Figure 2. The shear strength of unbroken seamounts, represented by the fracture strength of intact gabbro (Figure 9b), is much higher than the frictional strength of the surrounding sedimentary rocks (Figure 10a). Hence, if subducted seamounts remain unbroken, they would locally lock the plate boundary fault. This is the case for a subducted seamount that was observed under accreted sediments offshore of Shikoku, SW Japan; this seamount worked as a barrier against the rupture of the 1946 Nankaido earthquake (Kodaira et al.2000). Collapse of a seamount results in a local static stress drop (from fracture to frictional strength level) of about 80 MPa (shown as a dashed green arrow in Figure 10). It is possible that the breaking of an unruptured part of the seamount triggered the M9 Tohoku-oki earthquake. Strong-motion waveforms (Kumagai et al. 2012) and a simulation of seismic wave propagation (Duan 2012) support this idea. The average stress drop ($\Delta\tau$) over the faulted area of the Tohoku-oki earthquake was estimated to be 7 MPa by Lee et al. (2011), approximately 20 MPa by Hasegawa et al. (2011) and Yagi and Fukahata (2011), and 40 MPa by Kumagai et al. (2012) using seismic or geodetic methods. All these estimates are considerably larger than $\Delta\tau$~3 MPa of ordinary interplate earthquakes (Kanamori and Anderson 1975). The large local stress drop associated with breaking of a locked part of a seamount might have partly contributed to the observed large $\Delta\tau$.

Close relationships between subducted seamounts and the hypocenters of large earthquakes have been identified in many other subduction zones (e.g., Scholz and Small 1997), although the roles of seamounts in seismic rupture propagation were not the same for all the cases (Mochizuki et al.2008; Wang and Bilek 2011). At a shallow part of the accretionary wedge, overriding soft sediments easily deform and seamounts would behave as rigid objects (Dominguez et al. 1998) or barriers (Kodaira et al. 2000; Hirata et al. 2003). However, beneath the island-arc lower crust and mantle, there are no significant differences in yield strength between seamounts and overriding crystalline rocks (Figure 3b).

Wang and Bilek (2011) stated that faulting or 'decapitation' of seamounts is unlikely, but the common occurrence of fossil fragmented seamounts in outcrops of high-pressure metamorphic belts (e.g., Agata 1994; Terabayashi et al. 2005) indicates that breaking of seamounts did occur in ancient subduction zones.

Elastic strain accumulated in the frontal part of the accretionary wedge (segment o-a) would not have been large in the period preceding the Tohoku-oki M9 event because the imposed shear strain could be accommodated by the deformation of unconsolidated sediments. Clay-rich pelagic sediments that constitute the basal thrust of the accretionary wedge have velocity-strengthening frictional properties. Hence, the shallowest part near the trench axis is assigned to be a stably sliding zone. However, once the intermediate part (segment b-c) of the megathrust is broken, fault slip can easily propagate trenchward as demonstrated by the 'strong patch model' of Kato and Yoshida (2011). Because the frictional resistance is weaker at shallower levels due to the smaller overburden and the presence of clay-rich sediments (Figure 10a), fault movements would be accelerated to the point of inducing large tsunamigenic slips.

Kato and Yoshida (2011) assumed that the strength of interplate megathrusts was maximized at intermediate depths (approximately 20 km) close to the nucleation site of the M9 earthquake. The thrust fault on the down-dip side (>20 km) was thought to be weak as a result of increasing pore pressure (Yoshida and Kato 2011). The pore pressure distribution considered in Figure 6a is similar to Figure five(a) of Yoshida and Kato (2011), but the result of the model calculation using the parameters in Figure 9a shows that the shear strength was higher on the down-dip side of the M9 hypocenter because of the increasing importance of plastic deformation. The mechanism of rupture propagation to the down-dip side is discussed in the next section.

Most seamounts beneath the island-arc Moho would be faulted at their base, as the flexure force of the overriding plate resists the subduction of the seamounts (Cloos 1992; Scholz and Small 1997). Hence, the frictional plastic strength of gabbro was applied to the deep seamounts in Figure 10a. Quartz-rich sedimentary rocks become ductile at depths greater than 35 km, whereas gabbro is still brittle and strong at this depth. The deep regular asperity of the M7 Miyagi-oki earthquake is therefore considered to represent a subducted seamount, which was surrounded by sedimentary rocks. Geophysical evidence supports the existence of subducted seamounts at this depth as detailed above. The differences in shear strength and rheological behavior between quartz and gabbro offer an explanation for why deep earthquakes repeatedly occurred in such a limited area.

In contrast to the deep thrust zone, the shear strength of siliceous sedimentary rocks in the intermediate zone (segment b-c in Figure 6a) is almost the same as that of basic rocks. In addition, the difference in the friction coefficient between clay-rich and siliceous sedimentary rocks is not very large as the smectite-illite transition proceeds at this depth. The huge asperity associated with the M9 event can be understood as a consequence of the material-independent nature of the fault strength at intermediate depths. Heterogeneity within the M9 asperity, as represented by the distribution of small repeating earthquakes (Uchida and Matsuzawa 2011), and the slip areas of the 1981 Miyagi-oki earthquake (Figure 1) and the M7.3 foreshock event on 9 March 2011 (Ando and Imanishi 2011; Kato et al. 2012) may reflect physical conditions other than material properties, such as the topography of the fault surface and the distribution of H_2O fluids.

The depths of the M9- and M7-class asperities correspond to domains B and C of Lay et al. (2012), respectively (Figure 10a). Domain B is a source region for large-slip earthquakes characterized by large seismic regions (asperities), whereas isolated patches (asperities) in domain C generate modest-slip earthquakes. These domains correspond to the low- and high-frequency radiation zones of Tajima and Kennett (2012), respectively. The bimodal depth distributions of thrust-type earthquakes in subduction zones (Pacheco et al. 1993) possibly reflect these two types of asperities. The high shear strength of deep seamounts is consistent with the strong seismic wave radiation observed in domain C.

Depth Variations in Frictional Parameters

The stability of fault slip in the RSF law is described by the parameter a-b. At a high slip rate ($V > 10^{-2}$ m/s), several lubrication mechanisms such as frictional melting and thermal pressurization take place and a - b becomes negative irrespective of rock types (Di Toro et al. 2011). In contrast, a - b at low-to-intermediate slip rates varies with the experimental conditions and the materials involved. This section focuses on a-b at the low slip rates ($V \leq 10^{-6}$ m/s) that control earthquake nucleation processes. At low pressures ($n < 30$ MPa) and low to intermediate slip rates, dry quartz rocks and quartz gouges show a weak negative dependence of μ on V (i.e., a - b < 0) (Di Toro et al. 2011). Amorphization of quartz (Nakamura et al. 2012) is one of the possible velocity-weakening mechanisms under these conditions. Nakatani and Scholz (2004a) suggested neutral or negative a-b for the frictional law of quartz gouges that were healed under hydrothermal conditions (Pc = 60 MPa, T < 200°C). At temperatures above 300°C, wet quartz deforms plastically (Figure 9b). Obviously, the zone of fully plastic deformation is a stable region (a - b > 0). Shimamoto (1986) demonstrated for halite gouge that a transition from velocity-weakening to velocity-strengthening occurs within the brittle-ductile transition zone. Application of this result to the strength envelope of wet quartz yields a transition depth around 30 km (Figure 10b). The temperature of the plate interface at which a - b becomes neutral is about 200°C (Figure 6).

High-pressure ($\sigma'_n = 400$ MPa) experiments of granite gouge conducted at water-saturated conditions revealed a change in frictional behaviors from velocity-weakening (a - b < 0) at low-T (< 300°C) to strengthening (a - b > 0) at a temperature around 300°C (Blanpied et al. 1995). The frictional law of quartz gouge at high-pressure ($P_c = 250$ MPa and $P_p = 100$ MPa) hydrothermal conditions suggests that there is a similar change at lower temperatures between 150°C and 300°C (Chester and Higgs 1992; Chester 1995). The temperature range of the unstable-stable transition for wet quartz estimated above is consistent with these experimental results. At the depth of the M7-class asperity (35 to 45 km), a-b for quartz rocks becomes positive, whereas gabbro is brittle and likely to have a negative a-b (He et al. 2007). Therefore, a ruptured seamount surrounded by the siliceous sedimentary crust would behave in a manner typical of seismic asperities (Figure 10b); however, during the main shock event on 11 March 2011, both the M7-class asperity and its surroundings were ruptured. The initial 60 s of the event were associated with the fault slip to the up-dip side of the hypocenter; the rupture propagation to the down-dip side was observed 60 to 90 s after the initiation of rupturing (Koketsu et al. 2011; Yokota et al.2011). A possible explanation for the observed rupture propagation is as follows: siliceous sedimentary rocks surrounding the M7-class asperity were resistant to slip propagation in the initial 60 s because a-b was positive at the plate convergence rate (Figures 10b and 12a); then with increasing slip velocity, the unstable region of wet quartz expanded to the down-dip side (shown as an orange line in Figure 10b), and the outside of the M7-class asperity was converted to an unstable region. This was followed by a gradual acceleration in fault movement to seismic slip rates (Figure 12b,c). In contrast, if earthquakes are nucleated at the deep asperity, a drastic change in slip velocity is needed for the surrounding siliceous rocks to convert to velocity-weakening behavior. Hence, slip propagation would be prevented by the highly viscous resistance of the surroundings. As a result, fault movements at deep asperities cease within the asperities and do not evolve to generate gigantic earthquakes.

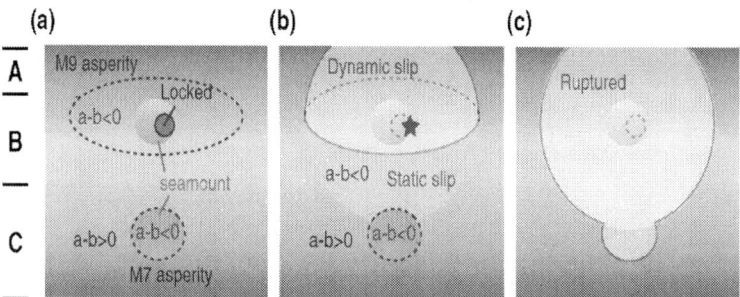

Figure 12: Frontal view of the interplate megathrust and possible scenario for slip propagation during gigantic earthquakes. A to C, rupture

domains defined by Lay et al. (2012). (a) Plate boundary structure during an interseismic period. The asperities associated with M9- and M7-class earthquakes are indicated by broken lines. A part of the M9 asperity is tightly locked by an unruptured seamount. The color gradation from red to blue represents the velocity dependence of shear strength. (b) Triggering of a gigantic earthquake (red star) by breakdown of the locked part of the seamount. The M9 asperity is broken by dynamic rupture. The slow slip region propagates downward. (c) Rupture propagation to the down-dip side of the megathrusts. See text for details.

In contrast to quartz-rich sediments, illite-rich gouges show velocity-hardening behaviors up to 250°C at $V = 10^{-6}$ m/s (den Hartog and Spiers 2013). Hence, the nucleation site of the M9 earthquake would not have been largely covered by clay-rich sedimentary rocks as mentioned above. However, the parameter a-b turned to negative in the temperature range of 250°C to 400°C, which corresponds to a depth of 35 to 60 km on the Tohoku-oki megathrust. Hence, the slow slip region with negative a-b (Figure 12b) can extend to clay-rich sedimentary rocks as well as siliceous rocks. Overall, the temperature range of the velocity-weakening regime varies with slip velocity, effective normal stress, the ratios of quartz and clays, and the compositions of clays or phyllosilicates (den Hartog and Spiers 2013; den Hartog et al. 2013).

Another important parameter involved in the constitutive equations of friction is the slip weakening distance D_c (Ohnaka 2003) or the length scales analogous to D_c in the RSF laws (herein, denoted as ' D_c' despite the small difference in definitions). Ohnaka (2003) and Perfettini et al. (2003) discussed the fact that D_c in natural fault systems is a scale-dependent parameter rather than a material property. If this is the case, any large asperities at intermediate depths would have a large D_c. This speculation is consistent with the results of the numerical simulations, which suggest that D_c in the M9 asperity needs to be much larger than those in the M7-class asperities (Hori and Miyazaki 2011; Kato and Yoshida 2011; Shibazaki et al. 2011).

The depth and material-dependent changes of , asperity size, a-b, and its velocity dependence are essential features of subduction zone megathrusts. Hence, incorporation of all these features will be needed to model earthquake cycles and rupture propagation during gigantic earthquakes in the future.

Megathrust Shear Strength and Influence of Pore Fluids

The absolute shear stress (τ_0) just before seismic events is equals to or slightly higher than the static strength τ during interseismic periods. Hasegawa et al. (2011) and Yagi and Fukahata (2011) discussed the fact that τ_0 on the large slip area of the M9 mainshock event was about the same as the stress drop $(\Delta\tau)$ in the same area, which was estimated to be about 20 MPa in their works. This means that the strength of the thrust fault before the earthquake was about 20 MPa, which is far smaller than the shear strengths estimated in the present rheological model (over 100 MPa in the hypocentral zone of the M9 event). τ_0 values of about 20 MPa were also obtained in heat flow (Furukawa and Uyeda 1989) and force balance models (Lamb 2006; Seno 2009; Wang et al. 2010) for the Tohoku-oki megathrust. The weakness of the thrust fault was thought to be a consequence of the extremely high pore pressure ratio $(\lambda \geq 0.95)$ at the plate interface (Seno2009). This section focuses on the absolute stress level and the role of high-pressure fluids on seismogenesis in the Tohoku-oki megathrust.

Yagi and Fukahata (2011) considered that the shear stress that accumulated on the plate interface was completely released during the M9 mainshock event because aftershocks in the coseismic slip area were predominantly the normal-fault type. However, in most of the focal mechanism data they analyzed (except for three data points), three were taken from the shallow (>10 km) parts of the accretionary prism. Asano et al. (2011) suggested that the focal mechanisms of the aftershocks that occurred at the shallow parts of the forearc region were different from those at the deeper parts.

Hasegawa et al. (2011) applied the near-field stress change model of Hardebeck and Hauksson (2001) to the Tohoku-oki megathrust and concluded that a nearly complete stress drop ($\Delta\tau/\tau_0 = 0.9$ to 0.95) occurred during the M9 mainshock event, although reverse-fault type events were still dominant for aftershocks at depths close to the plate interface. Their results must be interpreted carefully because the raw data of the principal stress axis directions (Figure three (a) and three(c) of Hasegawa et al. 2011) show considerably large scatter. Yang et al. (2013) showed that the stress states in the fore-arc crust was heterogeneous and depth dependent. Their estimates yield $_0$ values of approximately 40 MPa at depths of 5 to 15 km and 50 to 180 MPa at depths of 15 to 20 km. Chiba et al. (2012) examined the focal mechanism distributions before and after the M9 mainshock event in more detail and concluded that the principal stress directions were unchanged at the deepest part of the overriding plate. One possible explanation for the simultaneous occurrence of normal faulting within the accretionary wedge and reverse faulting along the basal thrust is the extrusion of the accretionary front, as shown in Figure one of McKenzie and Jackson (2012), although their calculation was made for a specific case of the stress-free basal thrust.

Furukawa and Uyeda (1989) expressed the relation between the heat production rate (q) at the plate boundary and the plate convergence rate as

$$q = \tau V_0$$

(15)

and derived $\tau = 10$ to 20 MPa for the seismogenic zone of the plate interface. Equation 15 implicitly assumes that the Pacific plate is stably sliding without slip deficits. However, the occurrence of the 2011 Tohoku-oki earthquake revealed that slip deficits had accumulated on the plate interface over a long interseismic period. The interplate coupling coefficient (c) before the earthquake was 0.5 to 0.8 on average (Uchida and Matsuzawa 2011), and is was close to 1 (i.e., back slip rate of approximately 80 mm/year) around the hypocenter of the M9 earthquake (Nishimura et al. 2000; Suwa et al. 2006). Modifying Equation 15 as

$$q = \tau(1 - c)V_0$$

(16)

and applying $c \geq 0.8$ to the large coseismic slip area, the apparent discrepancy between heat flow data and the frictional strength in Figure 10 is almost diminished.

Lamb (2006) assumed that the region behind the Tohoku-oki megathrust beneath the volcanic arc was in a neutral stress state (i.e., zero deviatric components in backstop stress). This assumption is inconsistent with the geodetic observations that show rapid compressional deformation of the Japanese Islands; rheological studies suggest that considerably high stress is required to deform the island-arc crust (Shimamoto 1993; Muto and Ohzono 2012). Ductile deformation in the aseismic part of the plate boundary was incorporated into the model, but simulated values of activation energy (36 to 37 kJ/mol[1]) were one order of magnitude less than that of mantle olivine. Seno (2009) employed a simple force balance model that makes no assumptions about the backstop stress and the failure criterions except for the frictional law at the base of accretionary prisms. The absolute value of τ was not exactly determined in this model but inferred from the focal mechanisms within the overriding plate; τ was set to positive (or negative) for the regions of compression (or tension). The distribution of τ in the forearc region offshore of Miyagi was poorly constrained because of the absence of neutral points. Wang et al. (2010) considered that the stress states in erosional margins are fluctuating during earthquake cycles and proposed a 'dynamic' Coulomb wedge model. Application of the dynamic Coulomb wedge model to the NE Japan subduction zone yields $\tilde{\lambda} > 0.95$ and a friction coefficient (τ/σ_v) of 0.03 in the hypocentral area (Kimura et al. 2012); however, the change of the stress field in the accretionary prism postulated in the model (i.e., extensional during interseismic periods but converts to compressive after a large earthquake) was the opposite of what was observed upon the M9 Tohoku-oki earthquake.

Using the classical Coulomb wedge theory (Davis et al. 1983), the pore pressure ratios at the base of the accretionary wedges can be derived without knowing the back stress distribution. The theoretical prediction is well supported by direct fluid pressure measurements in different types of subduction zones, although the hypothesis of the critical Coulomb failure states can be only justified in shallow parts of the forearc crust. It is noteworthy that the pore pressure ratio ($\tilde{\lambda}=0.5$) estimated for the Tohoku-oki megathrust is the smallest among 12 accretionary wedges analyzed by Davis et al. (1983). Geologic evidence of overpressurized fluids such as mud volcanoes has not been reported from the forearc region of NE Japan.

Generation of overpressurized fluids in subduction zones is related to several geological factors such as the permeability of water in the fault zone and wall rocks and volumes of incoming sediments (Saffer and Bekins 2002). In the case of erosional margins like NE Japan (Von Huene and Lallemand 1990), the net volume of incoming sediments would not be significantly large. Dehydration of hydrous minerals in subducting slabs is another important factor that can increaseP_p (Peacock et al. 2011; Saffer and Tobin 2011), but its influence would be moderate in cold subduction zones such as NE Japan because the specific volume of H_2O fluids along the plate boundary is small; for example, 0.9 to 1.0 cm³/g at 50°C to 150°C and 100 to 400 MPa according to Burnham et al. (1969).

Kimura et al. (2012) considered that opal-quartz transition in the pelagic sediments has led to the generation of high pore pressure fluids along the Tohoku-oki megathrust. In a strict sense, the opal-quartz transition is not a 'dehydration' reaction, as water is included in opal as H_2O molecules (not as hydroxyls). Whether or not this dewatering reaction generates excess pore pressure depends on the change in solid and fluid volumes and the permeability of the fault zone. Biogenic sediments consisting of amorphous or microcrystalline silica have large intergranular pore spaces, and additional pores are produced during transformation of low-density (approximately 2.1 g/cm³) particles of opal to denser (2.65 g/cm³) crystals of α-quartz:

$$SiO_2 \cdot nH_2O \text{ (Opal)} \rightarrow SiO_2 \text{ (Quartz)} + nH_2O \tag{17}$$

If 10 wt. % of the H_2O molecules contained in opal-A or opal-CT (Graetsch 1994) are totally released during the opal-quartz transition, then more than 28 vol. % of opal is converted to pores, whereas the liquid water produced is 21 vol. %. Because the net volume change is negative, overpressurized fluids are unlikely to be generated by silica diagenesis alone. A highly permeable zone produced by the opal-quartz transformation would become a fluid pathway. Hence, permeability along the oceanic sedimentary layer would increase and pore pressure would decrease. Similarly, generation of high pore pressure fluids by the smectite-illite transition (Kimura et al. 2012) is questionable because smectite contains water as H_2O molecules as in the case of opal.

Strong reflectors observed to the east of the M9 hypocenter (Kimura et al. 2012) and in the northern region offshore of Iwate (Fujie et al. 2002) suggest the presence of H_2O fluids along the plate boundary; however, the fluid pressures beneath the strong reflectors are not necessarily high. To date, low-V and high- V_p/V_s anomalies indicative of high-pressure fluids and serpentinites (e.g., Watanabe 2007; Peacock et al. 2011) have not been reported from the region offshore of Miyagi (Miura et al. 2005; Ito et al. 2005; Yamamoto et al. 2008). In contrast, the mantle wedge in the southern area (offshore of Fukushima) shows a low- $V_p (= 7.4$ km/s) (Miura et al. 2003; denoted by a blue bar in Figure 11) and a high- V_p/V_s anomaly (Yamamoto et al. 2008). Tomographic imaging of the hanging walls of the plate interface revealed domains of low-V and high- V_p/V_s anomalies to the north (offshore of Iwate) and to the south (offshore of Fukushima and Ibaraki) of the M9 hypocenter (Zhao et al. 2011; shown as light blue areas in Figure 11). Silent earthquakes have episodically occurred in the northern domain (Kawasaki et al. 2001). Tsuru et al. (2002) and Miura et al. (2003) identified high-porosity sedimentary units with values of $V_p = 2$ to 4 km/s at the base of the forearc crust along the trench axis (shown as brown bars in Figure 11). These porous sediments are widely distributed in the northern and southern parts of the Tohoku-oki megathrusts. These observations strongly suggest that the regions outside the M9 asperity are fluid-rich regions and that the interplate coupling was weakened by high-pressure fluids, serpentinite, and unconsolidated sediments in these regions.

CONCLUSIONS

The strength of the interplate megathrust that generated the 2011 Tohoku-oki earthquake was investigated based on the velocity structures offshore of Miyagi Prefecture and high-PT rheological properties of oceanic crustal materials. A new theoretical method was proposed to describe the change in shear strength in the brittle-ductile transition zone, in which the pore pressure increased to the lithostatic level. Noting the many simplifying assumptions made in constructing the model, and the uncertainties inherent in these assumptions, and in parameter values employed, the main results predicted by the model and their implications are as follows:

- The frictional strength at the M9 hypocentral zone was likely much higher than the clay-rich sediments in the along-trench zone. The large gradient in frictional strength on the trenchward side of the M9 hypocenter offers a viable explanation for the large tsunamigenic slips during the M9 event.

- Geophysical observations suggest that a subducted seamount existed at the hypocentral area of the M9 earthquake. A collapse of an unruptured part of the seamount would result in a local stress drop of about 80 MPa. This large stress drop is considered to be one of the possible causes of the gigantic earthquake.

- Available data imply that wet quartz shows plastic deformation at depths greater than 35 km, whereas gabbro is brittle and strong at the same depth. Thus, the M7-class asperity associated with the Miyagi-oki earthquakes is most likely a broken seamount that is surrounded by siliceous sedimentary rocks. The conditionally stable nature of the surroundings can be explained by the brittle-ductile transitional behavior of wet quartz.

- The shear strength of the thrust fault may be relatively

insensitive to rock types at intermediate depths (15 to 35 km in depth). Hence, large asperities can likely be formed in the intermediate zone of the interplate megathrusts. The M9 asperity on the Tohoku-oki megathrust occupied a fluid-poor region in the intermediate zone.

ACKNOWLEDGEMENTS

The author thanks K Otsuki for valuable comments and discussions, and CJ Spiers and an anonymous reviewer for helpful comments and constructive reviews. This work was supported by grants from the Japan Society for the Promotion of Science (KAKENHI No. 22340148, No. 70377985), by a Grant-in-Aid for Scientific Research on Innovative Areas (KAKENHI No. 21109995), and by the Observation and Research Program for Prediction of Earthquakes and Volcanic Eruptions from the Ministry of Education, Culture, Sports, Science and Technology (MEXT).

REFERENCES

1.   Agata T (1994) The Asama igneous complex, central Japan: an ultramafic–mafic layered intrusion in the Mikabu greenstone belt, Sambagawa metamorphic terrain. Lithos 33:241-263

2.   Ando R, Imanishi K (2011) Possibility of Mw 90 mainshock triggered by diffusional propagation of after-slip from Mw73 foreshock. Earth Planets Space 63:767-771 doi:105047/eps201105016

3.   Asano Y, Saito T, Ito Y, Shiomi K, Hirose H (2011) Spatial distribution and focal mechanisms of aftershocks of the 2011 off the Pacific coast of Tohoku Earthquake. Earth Planets Space 63:669-673 doi:105047/eps201106016

4.   Blanpied ML, Lockner DA, Byerlee JD (1991) Fault stability inferred from granite sliding experiments at hydrothermal conditions. Geophy Res Lett 18:609-612

faulted rock at high temperature and pressure. Tectonophysics 23:177-203

105. Stipp M, Stünitz H, Heilbronner R, Schmid SM (2002) The eastern Tonale fault zone: a 'natural laboratory' for crystal plastic deformation of quartz over a temperature range from 250 to 700°C. J Struct Geol 24:1861-1884

106. Suwa Y, Miura S, Hasegawa A, Sato T, Tachibana K (2006) Interplate coupling beneath NE Japan inferred from three-dimensional displacement field. J Geophys Res 111:B04402 doi:101029/2004JB003203

107. Tajima F, Kennett BLN (2012) Interlocking of heterogeneous plate coupling and aftershock area expansion pattern for the 2011 Tohoku-Oki Mw9 earthquake. Geophys Res Lett 39:L05307 doi:101029/2011GL050703

108. Takahashi M, Mizoguchi K, Kitamura K, Masuda K (2007) Effects of clay content on the frictional strength and fluid transport property of faults. J Geophys Res 112:B08206 doi:101029/2006JB004678

109. Takahashi M, Uehara S, Mizoguchi K, Shimizu I, Okazaki K, Masuda K (2011) On the transient response of serpentine (antigorite) gouge to stepwise changes in slip velocity under high-temperature conditions. J Geophys Res 116:B10405 doi:101029/2010JB008062

110. Terabayashi M, Okamoto K, Yamamoto H, Kaneko Y, Ota T, Maruyama S, Katayama I, Komiya T, Ishikawa A, Anma R, Ozawa H, Windley B, Liou JG (2005) Accretionary complex origin of the mafic-ultramafic bodies of the Sanbagawa belt, central Shikoku, Japan. Int Geol Rev 47:1058-1073

111. Tsuji T, Ito Y, Kido M, Osada Y, Fujimoto H, Ashi J, Matsuoka T (2011) Potential tsunamigenic faults of the 2011 off the Pacific coast of Tohoku Earthquake. Earth Planets Space 63:831-834

112. Tsuru T, Park J-O, Miura S, Kodaira S, Kido Y, Hayashi T (2002) Along-arc structural variation of the plate boundary at the Japan Trench margin: implication of interplate coupling. J Geophys Res 107(B12):2357 doi:101029/2001JB001664

113. Uchida N, Matsuzawa T (2011) Coupling coefficient, hierarchical structure, and earthquake cycle for the source area of the 2011 off the Pacific coast of Tohoku earthquake inferred from small repeating earthquake data. Earth Planets Space 63:675-679 doi:105047/eps201107006

114. Ujiie K, Tanaka H, Saito T, Tsutsumi A, Mori J, Kameda J, Brodsky EE, Chester FM, Eguchi N, Toczko S, Expedition 343 and 343T scientists (2013) Low coseismic shear stress on the Tohoku-Oki megathrust determined from laboratory experiments. Science 342:1211-1214 10.1126/science.1243485

115. Von Huene R, Lallemand S (1990) Tectonic erosion along the Japan and Peru convergent margins. Geol Soc Am Bull 102:704-720

116. Wada I, Wang K (2009) Common depth of slab-mantle decoupling: reconciling diversity and uniformity of subduction zones. Geochem Geophys Geosys 10:Q10009 doi:101029/2009GC002570

117. Wang K, Bilek SL (2011) Do subducting seamounts generate or stop large earthquakes. Geology 39:819-822 doi:10.1130/G31856.1

118. Wang K, Hu Y, von Huene R, Kukowski N (2010) Interplate earthquakes as a driver of shallow subduction erosion. Geology 38:431-434 doi:101130/G305971

119. Watanabe T (2007) Compressional and shear wave velocities of serpentinized peridotites up to 200 MPa. Earth Planets Space 2:233-244

120. Wong T-F, Biegel R (1985) Effects of pressure on the micromechanics of faulting in San Marcos gabbro. J Struct Geol 7:737-749

121. Yagi Y, Fukahata Y (2011) Rupture process of the 201 Tohoku-oki earthquake and absolute elastic strain release. Geophys Res Lett 38:L19307 doi:101029/2011GL048701

122. Yamamoto Y, Hino R, Suzuki K, Ito Y, Yamada T, Shinohara M, Kanazawa T, Aoki G, Tanaka M, Uehira K, Fujie G, Kaneda Y, Takanami T, Sato T (2008) Spatial heterogeneity

of the mantle wedge structure and interplate coupling in the NE Japan forearc region. Geophys Res Lett 35:L23304 doi:101029/2008GL036100

123. Yamanaka Y, Kikuchi K (2004) Asperity map along the subduction zone in northeastern Japan inferred from regional seismic data. J Geophys Res 109:B07307 doi:101029/2003JB002683

124. Yang YR, Johnson KM, Chuang RY (2013) Inversion for absolute deviatoric crustal stress using focal mechanisms and coseismic stress changes: the 2011 M9 Tohoku-oki, Japan, earthquake. J Geophys Res 118(B10):5516-5529 doi:10.1002/jgrb.50389

125. Yokota Y, Koketsu K, Fujii Y, Satake K, Sakai S, Shinohara M, Kanazawa T (2011) Joint inversion of strong motion, teleseismic, geodetic, and tsunami datasets for the rupture process of the 2011 Tohoku earthquake. Geophys Res Lett 38:L00G21 doi:101029/2011GL050098

126. Yoshida S, Kato N (2011) Pore pressure distribution along plate interface that causes a shallow asperity of the 2011 great Tohoku-oki earthquake. Geophys Res Lett 38:L00G13 doi:101029/2011GL048902

127. Yue H, Lay T (2011) Inversion of high-rate (1 sps) GPS data for rupture process of the 11 March 2011 Tohoku earthquake (Mw 9.1). Geophys Res Lett 38:L00G09 doi:101029/2011GL048700

128. Zhao D, Huang Z, Umino N, Hasegawa A, Kanamori H (2011) Structural heterogeneity in the megathrust zone and mechanism of the 2011 Tohoku-oki earthquake (Mw 9.0). Geophys Res Lett 38:L17308 doi:101029/2011GL048408

Chapter 5

Shape of Mole Nose Providing Minimum Axial Resistance

Yi Shen, Xuyan Hou, Yiwei Qin, Shengyuan Jiang, and Zongquan Deng

School of Mechatronics Engineering, Harbin Institute of Technology, West Dazhi Street, Harbin, 15001, China

ABSTRACT

Introduction

As a carrier of different sensors, moles can penetrate into the regolith automatically and keep investigating the subsurface environment continuously. In this section, features of several moles with different applications are introduced to explain why we choose a hammer-driven mole to study.

Mole Driven by a Hammer

In this section, the penetrating principle of a hammer-driven mole is illustrated and a circular arc shape for the front nose is proposed. Moreover, applying the penetrating principle, experiments of the mole with an arc-shaped nose are performed to observe the penetration phenomena in a simulated lunar regolith.

Mechanics Analysis

According to soil mechanics theory, regions of soil failure are divided and a mechanics model is established between soil and mole with an arc-shaped nose. The work is done to get approximate axial resistance equations which are analyzed with the defined geometric parameters caliber-radius-head.

EDEM Simulations

EDEM is leading global software based on discrete element method, whose main function is to analyze and observe the movement of particles. Lunar soil simulacrum is established to simulate axial resistance. Eventually, the theoretical results are validated by simulation.

BACKGROUND

As a novel technique of *in situ* investigation, a mole device employs a self-penetrating mechanism, which is considered to be a promising direction for space missions to get information about the geological structure, evolution, and physical and chemical properties of the material if we want to know whether there is life on Mars or the conditions were ever suitable for humans. It is more compact, is lightweight, and has low power consumption. Once it has penetrated into a certain depth, it can acquire geological information of the medium constantly, using various sensors.

In order to conduct deep space exploration, many countries have been involved in the development of a low-speed, unmanned sub-surface investigation device. Several prototypes of the mole have been designed and tested in laboratory conditions even though it started late. The mole was first developed by the Russian Federal Space Agency for the Mars-96 mission, called Mars 96 Penetrator, utilizing high speed to penetrate into the regolith, which can penetrate a deeper distance but cannot be used repeatedly and causes great damage to the detecting instruments, leading to data errors [1]. To overcome the disadvantages above, a hammer-driven mechanism was applied to the mole, which will be described in the next sections. In 2011, Japanese researchers proposed a robotic screw explorer which can excavate into soil and transport it backward automatically [2],[3]. Inspired by animals, such as mouse and earthworm, more and more researchers focus on the bionics design, e.g., an earthworm-type robot which can make use of the reactive force caused by pushing the discharged regolith above the robot [4],[5]. As for China, apart from the development of a thermal drill which is a combination of a rotary drill and a melting probe in Hong Kong Polytechnic University [6], it is seldom studied so far. In addition, it is worth mentioning that all those prototypes are still in the exploratory pilot phase and none is implemented in space mission successfully, which provides a great space for China to develop the automatic penetrating device.

Although discharging soil backward has great advantage compared with squeezing soil, the hammer-driven mole penetrates better on current technology. So this paper still focuses on optimization of a mole driven by a hammer. As opposed to that of the conventional rotary drill, forward motion of the mole is done by displacement and compression of the soil. Penetration depth depends on the matching of three qualities and shape of the front nose. So it is crucial to study the shape of the front nose. Up to now, the hammer-driven mole has several types of front nose, e.g., MMUM [7], derived from PLUTO developed by DLR for the Beagle 2 lander on the ESA Mars Express mission with added capability of sampling at the 60° front cone that can open during further penetration [8],[9]. Another interesting device is MUPUS on the Philae lander by PAS, whose front nose was designed with sharp and elastic barbs to get anchoring property [10]. Following the successful MUPUS development, a nonlinear conical shape (ogive-shaped tip angle starts from 45° at the base and 30° at the end) was applied to the KRET [11].

The remainder of this paper is organized as follows. The principle of operation of the hammer-driven mole is illustrated and an arc-shaped front nose is proposed in the 'Methods' section. Besides, experiment results of a convex arc-shaped nose in simulated lunar are also presented in the 'Methods' section. A mechanical model between the front nose and soil is built in the 'Results and discussion' section. What's more, EDEM simulations of the mole with different geometric parameters are given in the 'Results and discussion' section to further prove the relationship between axial resistance and caliber-radius-head, followed by the 'Conclusions' section.

METHODS

Principle of Penetration

As for the hammer-driven, low-speed type of mole, the basic principle of operation is to make full use of impacting movement to get energy needed for penetration. As shown in Figure 1, the mole mainly consists of seven parts: hammer, servo driving unit, casing, driving spring, buffer spring, escapement mechanism, and tether.

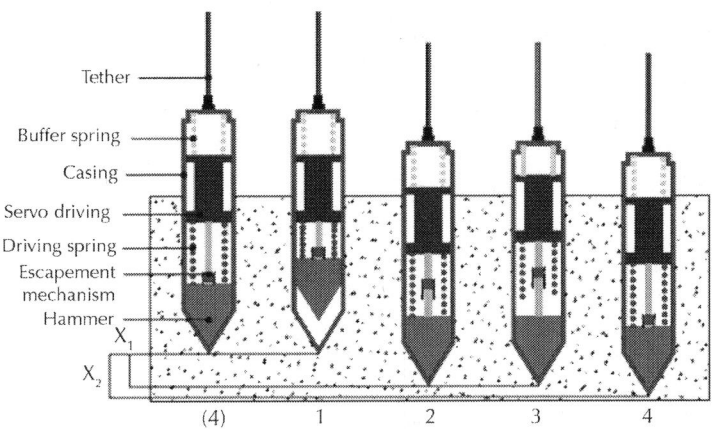

Figure 1: Schematic principle of the operation of the mole.

Therefore, the work cycle can be divided into four stages:

- Accumulation of energy in the driving spring by the movement of the hammer upward relatively to the casing via the servo driving unit.
- Escapement mechanism separates suddenly when the hammer reaches a certain distance and the driven hammer accelerates and hits the bottom of the casing, thus contributing the displacement x_1 by overcoming the restraint of the soil around.
- At the instant of the release of the hammer, the servo driving unit moves backward compared with the hammer, and then the buffer spring is compressed to avoid reverse movement of the mole.
- Under the effect of the buffer spring and its own gravity, the servo driving unit hits the casing, thus forcing the mole to move down at x_2. The movement of the mole repeats just like that, which is suitable for the regolith that can be compressed for making space. After the mole penetrates to a certain depth, a hole with a higher density appears at the back.

An Arc-shaped Front Nose

The shape and geometric parameters of the front nose, which contacts soil first, must have a great influence on the penetrating feature of the mole. Consider the axisymmetric mole as a rigid body, remaining undeformed during dynamic penetration. A convex arc shape of the front nose and notations are shown in Figure 2.

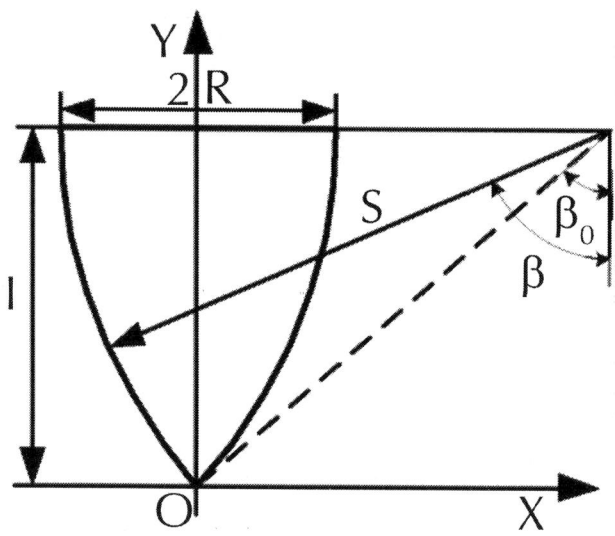

Figure 2: Cross section of a convex arc-shaped nose and notations.

The shape is arc of a circle with radius S that is tangent to the casing of the mole. Caliber-radius-head is defined as $=S/2R$, where R is the radius of the casing. The length of the front nose is l.

According to the geometric relationship, the following equations can be obtained:

$$
\begin{cases}
\sin\beta_0 = \dfrac{S-R}{S} \\
l = \sqrt{S^2-(S-R)^2} = R\sqrt{4\psi-1}
\end{cases}
$$

$$(1)$$

where K_p is the passive earth pressure coefficients and $K_p = tg^2(45° + \varphi/2)$.

Therefore, horizontal positive pressure on the infinitesimal surface can be expressed as:

$$dN = p_p dA = 2\pi p_p R dy \tag{22}$$

which can be integrated between $y = l$ and $(D - l)$ to give the net positive force N:

$$N = \int dN$$

$$= \pi R^2 \left[y \frac{K_P}{R} \left(D - R\sqrt{4\psi - 1} \right)^2 + 4c \frac{\sqrt{K_P}}{R} \left(D - R\sqrt{4\psi - 1} \right) \right] \tag{23}$$

Then we can get the resistance on the lateral surface:

$$F_c = \mu N$$

$$= \mu \pi R^2 \left[y \frac{K_P}{R} \left(D - R\sqrt{4\psi - 1} \right)^2 + 4c \frac{\sqrt{K_P}}{R} \left(D - R\sqrt{4\psi - 1} \right) \right] \tag{24}$$

Combining F_1' on the convex arc-shaped front nose, the axial force resisting motion of the mole can be obtained:

$$F_1 = F_1' + F_c \tag{25}$$

$$D > l + L$$

When the mole penetrates into the regolith totally, the length of the front nose is smaller than the penetration depth relatively and we can get the region of soil failure in Figure 10.

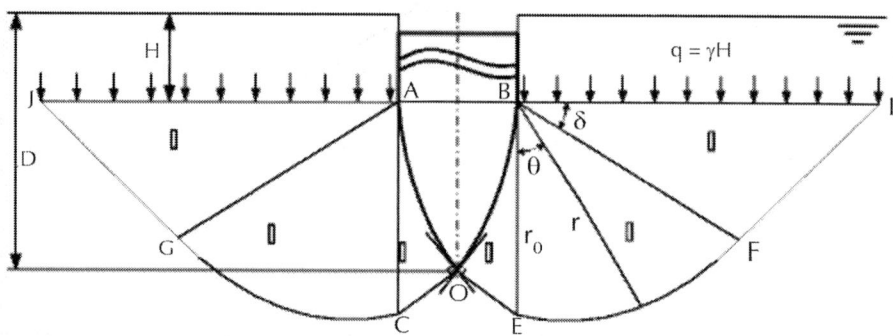

Figure 10: Soil failure when the mole penetrates totally.

Neglect the gravity of soil above the front nose and stress on surfaces AJ and BI can be regarded as the gravity of soil above the front nose. Then H is the distance from the ground to surfaces AJ and BI, which can be expressed as:

$$H = D-l = D-R\sqrt{4\psi-1}$$

(26)

Applying the mechanics analysis of soil block before, we can get the stress on surface BE:

$$p_F = \gamma N_q\left(D-R\sqrt{4\psi-1}\right) + cN_c$$

(27)

Afterwards, radical stress on the contact stress between soil and nose is:

$$p^{''} = a^{''}\sqrt{4\psi-1}\left(\frac{D}{R}-\sqrt{4\psi-1}\right) + cb^{''}$$

(28)

where

$$a^{''} = \frac{\gamma RN_q\left[2\tan\varphi\frac{\sqrt{4\psi-1}}{2\psi-1} + \left(1-\tan^2\varphi\right)\right]}{\left(1-\mu\tan\varphi\right)\sqrt{4\psi-1} + \left(\mu + \tan\varphi\right)}$$

$$b'' = \frac{2(N_c\tan\varphi + 1)\frac{4\psi-1}{2\psi-1} - [\tan\varphi - N_c(1-\tan^2\varphi)]\sqrt{4\psi-1}-1}{(1-\mu\tan\varphi)\sqrt{4\psi-1} + (\mu + \tan\varphi)}$$

According to Equation 19, integrate dF from $\theta = \beta_0$ and $\theta = \varpi/2$. So axial resistance on the front nose has the following expression:

$$F_1'' = \pi R^2\left[a''\sqrt{4\psi-1}\left(\frac{D}{R}-\sqrt{4\psi-1}\right) + cb''\right]$$

$$= \left[4\mu\left(\frac{\pi}{2}-\beta_0\right)\psi^2 - \mu(2\psi-1)\sqrt{4\psi-1} + 1\right] \quad (29)$$

Moreover, the lateral surface contacts soil totally in this stage. Combining $F_c = \mu N$, integrate Equation 22 between $y = l$ and $y = l + L$ to get the net resistance on the lateral surface:

$$F_c'' = \mu N = K_1' - K_2'\sqrt{4\psi-1} \quad (30)$$

where

$$K_1' = 2\mu L\sqrt{K_P}\left[\pi R y\sqrt{K_P}(2D-L) + c\right]$$

$$K_2' = 4\pi R^2\mu y K_P L$$

Then axial resistance of the mole is obtained as follows:

$$F_1 = F_1'' + F_c' \quad (31)$$

Axial Resistance vs Caliber-radius-head

Through mechanics analysis, it is known that for a convex arc-shaped front nose, axial resistance is identified with penetration depth and geometric parameters caliber-radius-head. In order to get the relationship between axial resistance and caliber-radius-head directly, other physical parameters which will be used are listed in Table 1 according to the acquaintance of the mole and soil.

Table 1: Physical parameters of the mole and soil

Physical parameters of the mole and soil						
	Parameters					
	R(mm)	L(mm)	c	φ(°)	γ(kN/m3)	μ
Value	15	200	0	30	18	0.3

Shen *et al.*

Shen *et al. Robotics and Biomimetics* 2014 1:10, doi: 10.1186/ s40638-014-0010-7

Based on axial resistance equations, curves relating axial resistance to penetration depth for various values of ψ are drawn (shown in Figure 11). As can be seen from Figure 11, the curve of axial resistance is similar to a parabola changing with penetration depth. Moreover, axial resistance decreases with the increase of ψ. However, the advantage is not obvious when ψ increases to 3 and a very large value of ψ causes an increase in total length of the front nose which is not advisable.

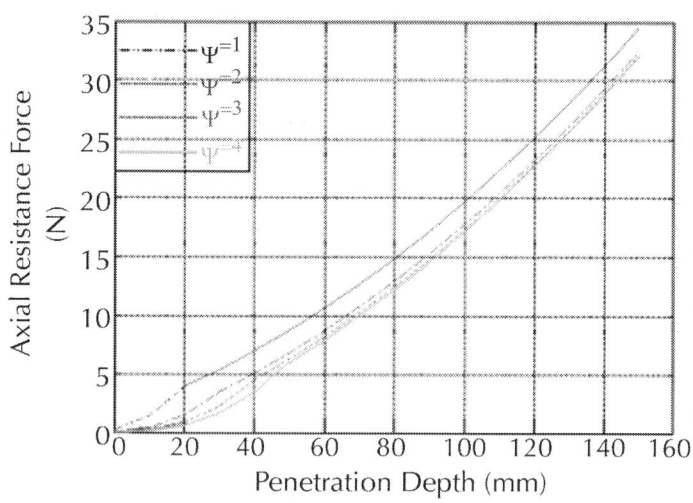

Figure 11: Axial resistance vs penetration depth for differentψ.

EDEM Simulation

Generally, granular matter behaves like a compressible non-Newtonian complex fluid including fluid solid transition and can be simulated using EDEM simulation. An obvious advantage of EDEM simulation is that it provides the possibility of obtaining movement, forces, and other dynamical properties of the system at any time.

For ensuring establishment of the simulation model consistent with theoretical analysis, the model of the mole is built in SolidWorks and has the same parameter setting as shown Table 1. Besides, the Hertz-Mindlin (no slip) interaction model is employed in EDEM simulations, and the setting of the other variables is shown in Tables 2 and 3. Simulation of the mole whose shape parameter is 2 is shown in Figure 12 to observe the flow of the lunar soil.

Table 2: Material parameters

Material	Poisson's ratio	Shear modulus (Pa)	Density (kg/m3)
Particle	0.3	2×10^7	1,750
Mole	0.3	2×10^{10}	7,850

Shen et al.

Shen et al. Robotics and Biomimetics 2014 1:10, doi: 10.1186/s40638-014-0010-7

Table 3: Interaction parameters

Interaction	Particle-particle	Particle-mole
Coefficient of restitution	0.1	0.2
Coefficient of static friction	0.574	0.3
Coefficient of rolling friction	0.01	0.01

Shen et al.

Shen et al. Robotics and Biomimetics 2014 **1**:10, doi: 10.1186/s40638-014-0010-7

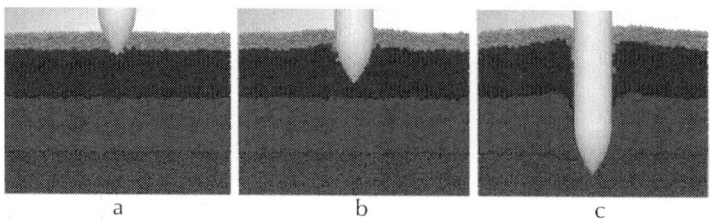

a   b   c

Figure 12: EDEM simulation of a mole with an arc-shaped front nose (a-c). The shape parameter of the mole is 2(ψ =2), and particles are set with different colors by section to observe the flow directly during the penetration.

It can be seen that the mole works properly in granular matter and the failure of soil increases along with the increase of penetration depth. Change of the values of the shape parameter ψ from 1 to 4 and all simulations are shown in Figure 13.

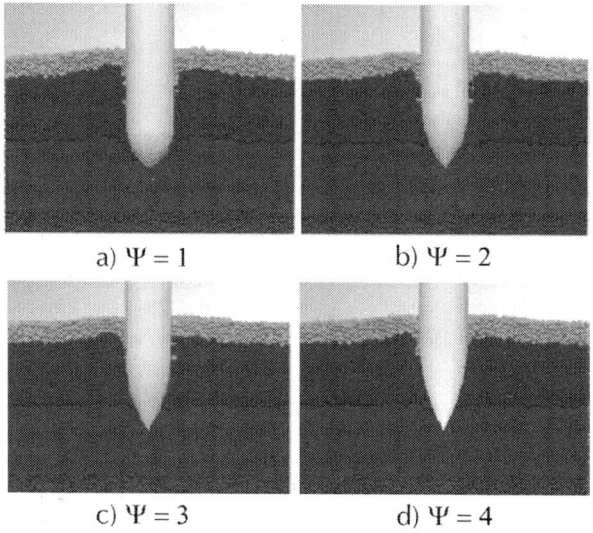

a) $\Psi = 1$ b) $\Psi = 2$

c) $\Psi = 3$ d) $\Psi = 4$

Figure 13: Penetration simulations of mole with an arc-shaped front nose for different ψ. (a)ψ =1. (b)ψ =2. (c)ψ =3. (d)ψ =4.

According to Figure 13, it can been seen that when the mole penetrates into the same depth, there is vertical displacement of

Thermal Effect on Wellbore Stability during Drilling Operation with Long Horizontal Section

Mengbo Li[a], Gonghui Liu[a, b], and Jun Li[a]

[a]China University of Petroleum, College of Petroleum Engineering, Beijing 102249, China

[b]Beijing Information Science & Technology University, Beijing 100192, China

ABSTRACT

In this study, a three-dimensional wellbore stability model is presented that takes into account thermal stresses combined with an integrated circulation temperature model for horizontal well drilling, the bottom hole temperature simulation were then validated using field measurements, and compared with results for vertical wells. A subsequent analysis of temperature sensitivity revealed that the heat source term, the length of horizontal section and mud specific heat were the main reasons cause the bottom hole temperature for horizontal wells rises above the static formation temperature. Results from the wellbore stability model show that the temperature variation magnitude in horizontal well is smaller than that in vertical wells, however, the effect of thermal stress on critical mud weight window in horizontal is more sensitive. The wellbore at toe of horizontal section is more stable than that at heel of horizontal section when the bottom hole temperature exceeds the static formation temperature. This research can provide a theoretical reference for enhancing overall operational efficiency and safety for horizontal well drilling.

INTRODUCTION

In oil and gas resource development and exploitation, horizontal wells have been widely used to enhance production by increasing the amount of wellbore that has contact with the target reservoir. Unfortunately, due to high temperature environments that accompany such drilling, severe problems can be encountered such as the instabilities caused by drilling fluids and in-situ stresses.

Many scholars have studied the heat transfer of geothermal systems during drilling operations, using both analytical and numerical methods to estimate the circulating fluid temperature. Analytical methods are generally used for modeling simple drilling systems, for example, with regular wellbore geometry and single temperature gradients (Holmes and Swift, 1970, Ramey, 1962, Edwardson et al., 1962, Tragesser et al., 1967 and Kabir et al., 1996). For more complex systems, however, simple analytical methods are unable to accurately model the thermal behavior. Numerical methods are required for studying more complex systems, and to provide a powerful predictive tool that can efficiently solve the governing finite difference equations for unsteady-state heat transfer in both wellbore and formation (Raymond, 1969,Wooley, 1980 and Marshall and Bentsen, 1982).

All mentioned above was suit for vertical wells, and some scholars (Perkins and Gonzalez, 1984 and Lin and Pinya, 1998) presented model predictions of the effect on the near wellbore stresses of different temperature for vertical wells.

More recently, as horizontal well have been widely used, Yoshioka et al. (2007) and Li and Zhu (2010)developed thermal models to predict downhole temperature, pressure and flow rate profiles for horizontal wells, but these models only consider the heat transfer in horizontal section and the reservoir. Kumar et al., 2012a and Kumar et al., 2012b developed a simple analytical model to analyze heat generated from borehole friction and to predict downhole temperatures for extended-reach well drilling operations. Their model applies only to steady-state conditions and therefore does not accurately model the heat transfer processes. Iyoho et al. (2009) discussed the influencing factors on wellbore temperature of horizontal wells with long horizontal or near horizontal sections for mud system design purposes, no theoretical details were revealed. Gonzalez et al. (2004) found that the fracture gradient can be influenced by wellbore temperature through leak-off test. Yu et al. (2001) and Nguyen et al. (2010) modeling the thermal effects on wellbore stability, separately. However, only limited studies presently exist that use numerical methods to study the thermal effect on wellbore stability combining with thermal behavior of horizontal well drilling systems, especially with long horizontal section.

The temperature variation in horizontal wells is much different from vertical wells (Trichel and Fabian, 2011), which cause a different thermal effect on wellbore stability along the long horizontal section, thus, we can't study either the temperature model or thermal effect on wellbore stability independently.

Given the above, the objective of our present research was to develop a combined model that would serve to: (i) numerically simulate the heat transfer processes during high temperature drilling operations in horizontal wells, and (ii) determine the thermal effect on wellbore stability under the true downhole drilling environment with long horizontal wells, based on temperature distribution derived from the simulations. In this study, an integrated circulation temperature model of horizontal well drilling was established to investigate the heat transfer characteristics of horizontal wells. Thermal stress near wellbore of horizontal well were analyzed combining with the true downhole drilling environment, the thermal effects on the "critical mud weight window" was discussed, providing a theoretical reference for better understanding the thermal behavior and thermal effect on wellbore stability in horizontal well drilling operations.

WELLBORE TEMPERATURE OF HORIZONTAL WELL AND INFLUENCING FACTORS

Mathematical Model Development

A schematic diagram of the horizontal drilling operation is shown in Fig. 1. The whole drilling system has five distinct regions: (1) drilling fluid flow downward through the drill pipe; (2) drill pipe wall region; (3) drilling fluid flow upward through the annular; (4) formation region, and (5) drill bit region. According to the well trajectory, each region can be divided into three parts: vertical section, curved section, and horizontal section (excluding region (5) above).

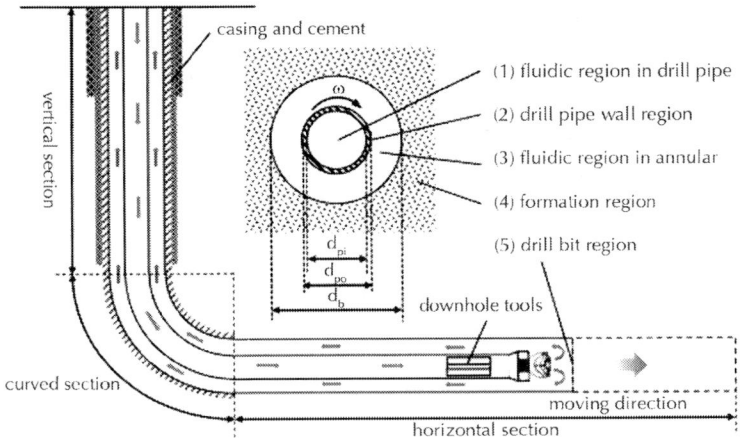

Figure 1: Illustrative sketch of fluid circulation in horizontal well drilling systems.

To develop the energy equations for describing thermal behavior of the entire wellbore profile and surrounding formation, the following assumptions are made:

- Only heat conduction in horizontal direction is considered, as the majority of formations drilled are layered rock.

- Physical properties of formations, i.e., density, specific heat and thermal conductivity rate are constant; heat conduction only is applied in modeling the formations.

- Fluid properties are independent of temperature, and wellbore drilling fluids are incompressible and in steady-state flow during each time step.

- Heat transfer within the drilling fluid occurs by axial convection. Conduction is neglected except when the circulation process is terminated.

- Horizontal well drilling has a rotational motion drilling, without any buckling.

In applying these equations to model the thermal behavior of the entire drilling system, five different sets of governing differential equations must be defined, one for each of the five regions identified earlier, along with boundary conditions at each interface as determined by flow continuity or other conditions.

The Wellbore Region

The wellbore region can be divided into three sub-regions which are fluidic region in drill pipe, drill pipe wall region and fluidic region in annular. The energy conservation equations in each sub-region in cylindrical coordinates can be written as the following forms, separately:

$$(\rho C_P)_{flui}\left(\frac{\partial T_{flui,p}}{\partial t} + v_{flui,p}\frac{\partial T_{flui,p}}{\partial z}\right) = \lambda_{flui}\frac{\partial^2 T_{flui,p}}{\partial r^2} + \frac{\lambda_{flui}}{r}\frac{\partial T_{flui,p}}{\partial r} + S_p \tag{1}$$

$$(\rho C_P)_{ste}\frac{\partial T_{pipe}}{\partial t} = \lambda_{ste}\frac{\partial^2 T_{pipe}}{\partial z^2} + \lambda_{ste}\frac{\partial^2 T_{pipe}}{\partial r^2} + \frac{\lambda_{ste}}{r}\frac{\partial T_{pipe}}{\partial r} \tag{2}$$

$$\left(\rho C^P\right)_{mix}\left(\frac{\partial T_{flui,a}}{\partial t} + v_{flui,a}\frac{\partial T_{flui,a}}{\partial z}\right)$$
$$= \lambda_{mix}\frac{\partial^2 T_{flui,a}}{\partial r^2} + \frac{\lambda_{mix}}{r}\frac{\partial T_{flui,a}}{\partial r} + S_a \tag{3}$$

Formation Region

The formation encountered in well drilling operations is porous medium. Formation porosity must therefore be considered in this region. The formation thermal capacity can be written as (Qiu et al., 2004):

$$(\rho C_P)_{form} = (\rho C_P)_{rock}(1 - \Phi) + (\rho C_P)_{poro}\Phi \tag{4}$$

$$\lambda_{form} = \lambda_{poro}^{\Phi} + \lambda_{rock}^{(1-\Phi)} \tag{5}$$

The energy balance equation for the formation regions in vertical and curved section of the wellbore can be written as:

$$(\rho C_P)_{form}\frac{\partial T_{form}}{\partial t} = \lambda_{form}\frac{\partial^2 T_{form}}{\partial r^2} + \frac{\lambda_{form}}{r}\frac{\partial T_{form}}{\partial r}$$

(6)

In horizontal drilling section, the axis of the wellbore is parallel with the horizontal direction. Heat conduction both in the axis and radial directions must therefore be considered. Furui et al. (2003) investigated a reservoir inflow model for a horizontal well and approximated the pressure and temperature profile in the reservoir as a composite of 1D radial flow near the well, and a 1D linear flow further away from the well as shown in Fig. 2 below. They estimated that the distance from the wellbore where linear streamlines became radial as H/2, where H is as shown in the Figure.

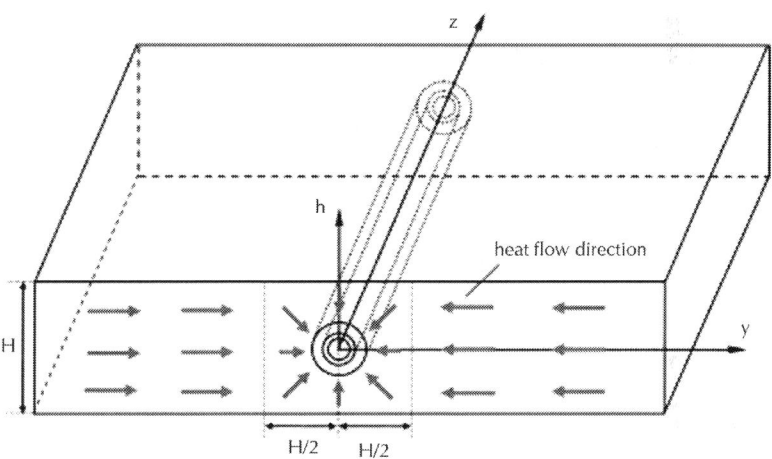

Figure 2: Schematics of heat transfer between wellbore and formation in horizontal section.

When there is no fluid flow in the formation (the region of linear streamlines), only heat conduction is considered, the energy balance equation then reduces to:

$$(\rho C_P)_{form} \frac{\partial T_{form}}{\partial t} = \begin{cases} \lambda_{form} \dfrac{\partial^2 T_{form}}{\partial r^2} + \dfrac{\lambda_{form}}{r} \dfrac{\partial T_{form}}{\partial r} + \lambda_{form} \dfrac{\partial^2 T_{form}}{\partial z^2}, & r \leq H/2 \\ \lambda_{form} \dfrac{\partial^2 T_{form}}{\partial r^2} + \lambda_{form} \dfrac{\partial^2 T_{form}}{\partial z^2}, & r > H/2 \end{cases}$$

$$(7)$$

Model Validation

A deep horizontal well (vertical depth 6180 m) in Tarim oilfield, with a long horizontal section (678 m) was selected with actual drilling circulation schedule. This well utilized an 88.9 mm diameter drill pipe set in a 152.4 mm diameter hole, with a very high bottom hole static temperature in the horizontal section of about 150 °C. Oil-based drilling fluid was used for best borehole stability. Rate of penetration varied from 0.71 to 3.76 m/h, with an average rate of 1.6 m/h. The physical properties of materials in drilling circulation system are shown in Table 1.

Table 1: Physical properties of materials in drilling circulation system

	Density [kg/m³]	Specific heat [J/(kg °C)]	Thermal conductivity [W/(m °C)]
Rock	2640	837	2.25
Drilling pipe	7800	400	43.75
Cement	1900	2000	1.0
Drilling fluid	1080	1647 (oil based mud)	1.02

MWD (measurement while drilling) was used to detect downhole annular temperature in both wells. The measured temperature was stored in the downhole tools. The device was able to continuously record the downhole temperature in the annulus during circulation.

Fig. 3 plots the predicted and measured temperatures at different depths during drilling operations. Time-zero was set to coincide with the time that the moving interface was at the landing point (the beginning of the horizontal section). The static temperature profile was used as the starting temperature at the commencement of the simulation. This was not optimal, but after 8 h of simulation, results were seen to be not significantly dependent on the initial conditions any longer. The calculated temperatures were close to measured values, and the bottom hole temperature gradually increased with the measured depth increase, as the drill bit continued to break more rock.

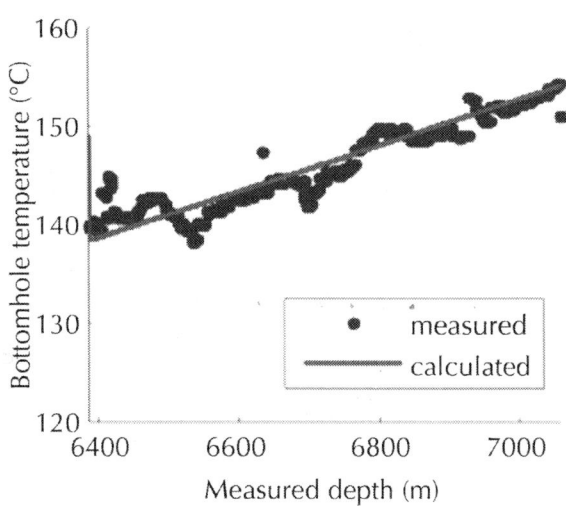

Figure 3: Trend match of bottom hole temperature measurements with model predictions.

Influencing Factors on Wellbore Temperature of Horizontal Well

For vertical wells, the change of wellbore temperature was mainly due to the heat transfer between the formation and wellbore fluid. Because of geothermal gradients, formation temperature varies greatly in the vertical direction. Compared with the heat change caused by geothermal gradients, minor thermal effects such as the heat caused by viscous dissipation in the drilling fluid can be ignored. However, for horizontal wells, the temperature gradient rarely changes in the horizontal section. When sufficient heat transfer occurs between the long horizontal section of the wellbore and the formation, any thermal effect will have a considerable impact on the changes in wellbore temperature.

Fig. 4 shows the calculated heat flux distributions attributable to forced convection, hydraulic energy and mechanical energy hear sources, for various unit control volumes at different sections of the wellbore in horizontal well drilling. The heat transfer and energy source distribution ratios are calculated at the midpoints of the respective unit control volumes for the vertical section, curved section, horizontal section, and the drill bit section. From Fig. 4, it is apparent that as the depth increases, the transferred heat caused by forced convection in unit control volumes decreases, mainly because the horizontal section is in full contact with the formation, which leads to small changes of fluid temperature in the axial direction. However, the heat generated by hydraulic energy and mechanical energy gradually increases with depth, and displays a significant increase near the drill bit. The heat transfer between the formation and the horizontal section of the wellbore, and mechanical and heat sources themselves are the main reasons for the different (i.e., higher) bottom hole circulating temperatures in horizontal wells versus vertical wells.

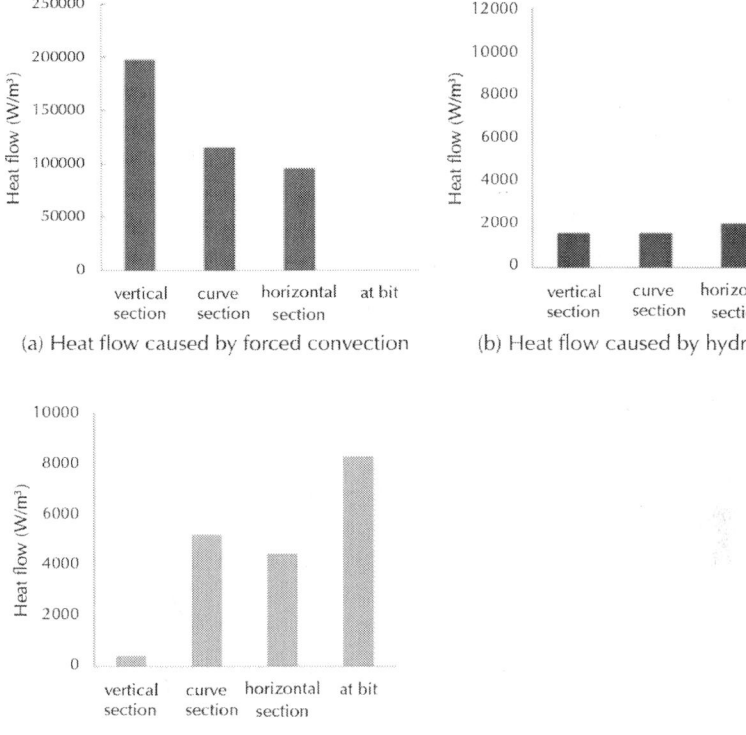

(a) Heat flow caused by forced convection

(b) Heat flow caused by hydraulic energy

(c) Heat flow caused by mechanical energy

Figure 4: Comparison of control volume heat flux in different sections of the wellbore.

The horizontal well geometry and materials physical properties were used as a basis for a sensitivity analysis. The major difference between horizontal wells and vertical wells in temperature profile is that the horizontal section of the wellbore has a sufficient heat transfer with the formations, where the bottom hole static temperature is at its maximum and the temperature gradient rarely changes. The length of the horizontal section reflects the size of the contact surface between the drilling fluid and the formation, as shown in Fig. 5. The bottom hole circulating temperature increases with length of the horizontal section. When this length reaches 700 m or more, the bottom hole circulating temperature exceeds the static formation temperature.

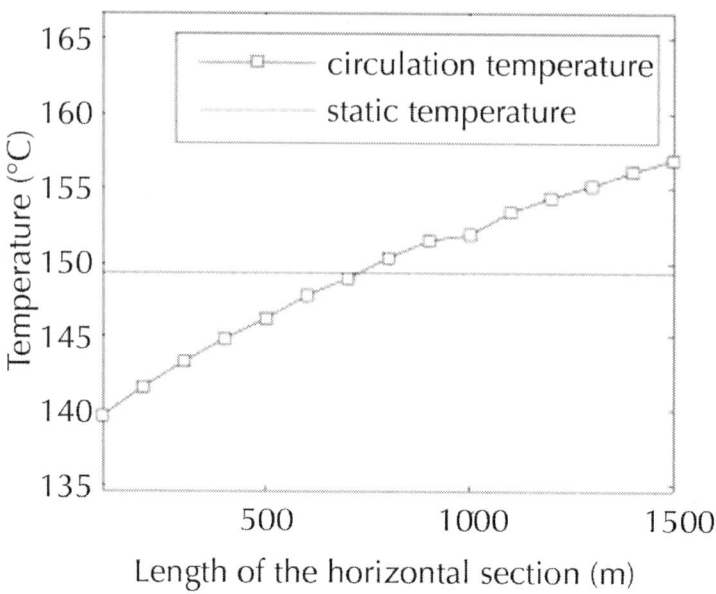

Figure 5: Effect of horizontal section length on the annular temperature profile.

Fig. 6 shows the effect of pressure drop distribution on the wellbore circulation temperature profile. The wellbore circulation temperature profiles of the horizontal section are affected mostly by the pressure drop, which causes an 8 °C increase compared with the bottom hole circulation temperature, without considering the frictional pressure drop in BHA. Fig. 7 shows the effect of the heat source component generated by torque on the annular temperature profile. There is a 3 °C decrease in bottom hole calculation temperature when without considering heat source term of torque.

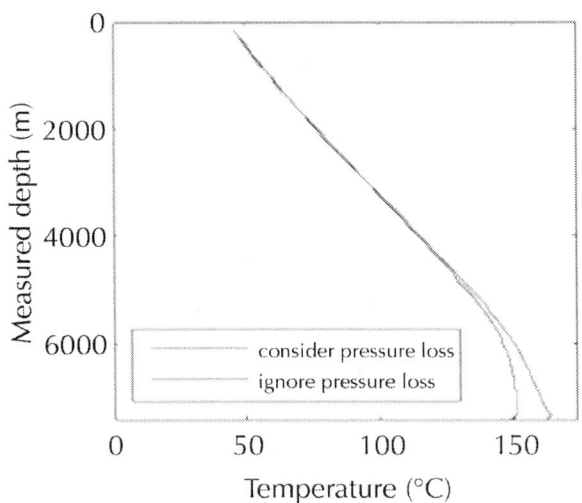

Figure 6: Effect of BHA pressure drop on the annular temperature profile.

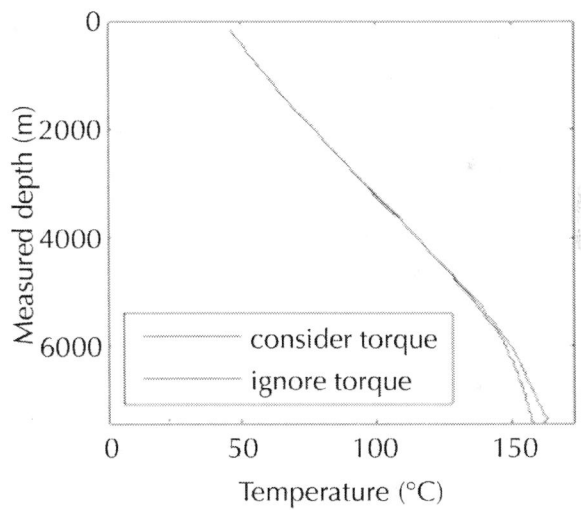

Figure 7: Effect of torque on the annular temperature profile.

Fluid systems with different specific heats generally exhibit marked differences in their respective thermal behavior. Fig. 8

demonstrates the effect of different mud systems on borehole circulation temperature profile. The specific heat capacity of water-based mud is about twice that for oil-based mud, resulting in the small temperature gradient of water-based mud in the annular temperature profile. Compared with oil-based mud, the annular temperature profile of water-based mud has a 20 °C decrease in its bottom temperature, and a 7 °C increase in outlet temperature.

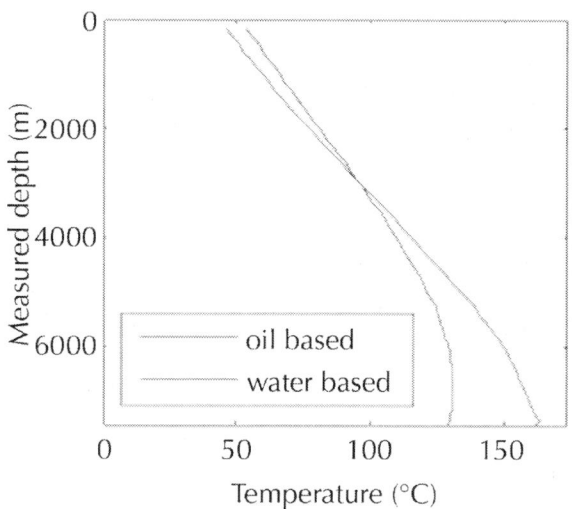

Figure 8: Effect of mud type (oil-based versus water-based) on annular temperature profile.

Reducing mud pump rate, on the one hand, will reduce convective heat transfer coefficient between the drilling fluid and formation, which in turn results in increased bottom hole circulating temperature. On the other hand, it reduces the pressure loss heat source term generated by the downhole tool, decreasing the bottom hole circulation temperature. Consequently, there is no obvious law between pump rate and bottom hole temperature, and the relationship should be determined according to the specific situation.

THERMAL STRESS NEAR WELL-BORE OF HORIZONTAL WELL

Stress Induced by Formation Temperature Changes in Horizontal Well

When the temperature near the wellbore is changed, the thermal stress occurs in the formation. It is well known that the thermal stress occurs only when the heating expansion or cooling shrinkage is restrained. The thermal stress resulting from a temperature change T (T-T0) for the rock which is fully constrained in one direction is:

$$\sigma_T = E\alpha_T \Delta T$$

(8)

where E is Young's modulus and α_T is linear thermal expansion co-efficient of rock. If the ΔT is positive, it will cause a tensile stress, and it has an opposite effect when the ΔT is negative.

The stresses induced by thermal effect near wellbore are given as follows:

$$\sigma_r = -\frac{E\alpha_T}{1-v}\frac{1}{r^2}\int_R^r \Delta T r dr$$

$$\sigma_\theta = \frac{E\alpha_T}{1-v}\frac{1}{r^2}\left[\int_R^r \Delta T r dr - r^2 \Delta T\right]$$

$$\tau_{r\theta} = -\frac{E\alpha_T}{1-v}\Delta T$$

(9)

A homogeneous and isotropic formation around the wellbore is assumed with constant formation properties. Based on the thermal elastic mechanics theory (Xu, 1982), the thermal effect is considered by adding the thermal induced stress into the pure elastic model

(Fjaer et al., 1992), and the stress components can be expressed in cylindrical coordinate (align with the wellbore axes) as follows:

$$
\begin{cases}
\sigma_r = \frac{R^2}{r^2}p_i + \frac{(\sigma_{xx}+\sigma_{yy})}{2}\left(1-\frac{R^2}{r^2}\right) + \frac{(\sigma_{xx}-\sigma_{yy})}{2}\left(1+\frac{3R^4}{r^4}-\frac{4R^2}{r^2}\right)\cos 2\theta + \delta\left[\frac{\alpha_B(1-2v)}{2(1-v)}\left(1-\frac{R^2}{r^2}\right)-\phi\right](p_i-P_p) - \frac{E\alpha_T}{1-v}\frac{1}{r^2}\int_R^r \Delta T r\,dr \\[2mm]
\sigma_\theta = -\frac{R^2}{r^2}p_i + \frac{(\sigma_{xx}+\sigma_{yy})}{2}\left(1+\frac{R^2}{r^2}\right) - \frac{(\sigma_{xx}-\sigma_{yy})}{2}\left(1+\frac{3R^4}{r^4}\right)\cos 2\theta + \delta\left[\frac{\alpha_B(1-2v)}{2(1-v)}\left(1+\frac{R^2}{r^2}\right)-\phi\right](p_i-P_p) + \frac{E\alpha_T}{1-v}\frac{1}{r^2}\left[\int_R^r \Delta T r\,dr - r^2\Delta T\right] \\[2mm]
\sigma_z = \sigma_{zz} + v\left[\sigma_{xx}+\sigma_{yy}-2(\sigma_{xx}-\sigma_{yy})\left(\frac{R}{r}\right)^2\cos 2\theta\right] + \delta\left[\frac{\alpha_B(1-2v)}{1-v}-\phi\right](p_i-P_p) - \frac{E\alpha_T}{1-v}\Delta T
\end{cases}
$$

$$(10)$$

where, the σ_{xx}, σ_{yy}, σ_{zz} shown above are the three in-situ principal stresses around the wellbore, for horizontal well, they have the following equations:

$$
\begin{cases}
\sigma_{xx} = \sigma_v \\
\sigma_{yy} = \sigma_H \sin a^2\beta + \sigma_h\cos^2\beta \\
\sigma_{zz} = \sigma_H\cos^2\beta + \sigma_h\sin^2\beta
\end{cases}
$$

$$(11)$$

At the borehole surface, the equation (10) can be express as follows:

$$
\begin{cases}
\sigma_r = p_i - \delta\phi(p_i - P_p) \\[2mm]
\sigma_\theta = -p_i + (1 - 2\cos 2\theta)\sigma_{xx} + (1 + 2\cos 2\theta)\sigma_{yy} + \delta\left[\frac{\alpha_B(1-2v)}{(1-v)}-\phi\right](p_i-P_p) - \frac{E\alpha_T}{1-v}\Delta T \\[2mm]
\sigma_z = \sigma_{zz} - 2v(\sigma_{xx}-\sigma_{yy})\cos 2\theta + \delta\left[\frac{\alpha_B(1-2v)}{1-v}-\phi\right](p_i-P_p) - \frac{E\alpha_T}{1-v}\Delta T
\end{cases}
$$

$$(12)$$

Failure of the Wellbore

The stresses around the wellbore are then calculated based on the pore pressure and temperature profiles by using the equations shown above. Critical mud weights are determined using Mohr–Coulomb failure criteria and tensile failure criteria,

The Mohr–Coulomb failure criteria has the following forms:

$$
(\sigma_{max} - \alpha P_p) \le 2S_0 \tan\left(\frac{\pi + 2\phi}{4}\right) + (\sigma_{min} - \alpha P_p)\tan^2\left(\frac{\pi + 2\phi}{4}\right)
$$

$$(13)$$

failure criteria, the effects of Poisson's ratio must be considered. Aadnoy and Belayneh (2008) described the effects of Poisson's ratio on fracturing pressure and thermal stress by using the scaling factor

C and K, the temperature effect on the fracturing equation can be expressed as:

$$\sigma_T = KE\alpha_T \Delta T$$

(14)

where, K = $(1 + \upsilon)^2/(3\upsilon(1 - 2\upsilon) + (1 + \upsilon)^2)$, Fig. 9 shows the magnitude of Poisson's effect.

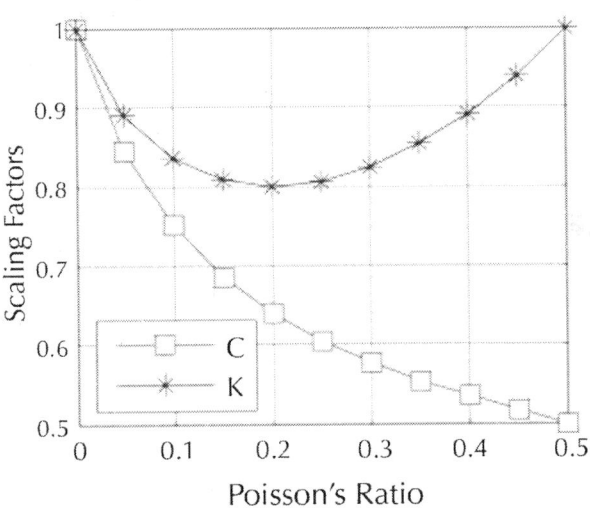

Figure 9: Scaling factors caused by Poisson's effect (Aadnoy and Belayneh, 2008).

And the tensile failure criteria can be expressed as:

$$\sigma_{bd} = \left(\sigma_{min} - \alpha P_p\right) + \sigma_t \leq 0$$

(15)

The critical mud weight window are computed by use of equations above, the input parameters are given in Table 2 including thermal effects, wellbore information. The examples shown in this paper were based on this input data combining with well geometry and physical properties of materials discussed in Section 2.2.

Table 2: Input data of formation properties

Variables	Values
Overburden stress gradient	2.26 Mpa/100 m
Maximum horizontal stress gradient	2.04 MPa/100 m
Minimum horizontal stress gradient	1.88 Mpa/100 m
Pore pressure, equivalent	1.24 g/cm³
Thermal expansion coefficient	$2.36 \times 10^{-5}\ °C^{-1}$
Poisson's ratio	0.22
Biot's constant	0.9
Young's modulus	6895 Mpa
Cohesion	6.14 Mpa
Friction angle	30°
Tensile strength	0.69 Mpa

The minimum and maximum mud weight requirements for both breakdown and collapse as a function of borehole inclinations are plotted, as shown in Fig. 10. As the inclination angle increased, the mud weight for breakdown decreased, and mud weight for collapse increased, which makes the safety mud window of horizontal well narrower than that in vertical well, and the horizontal wellbore is more apt to fracture or collapse.

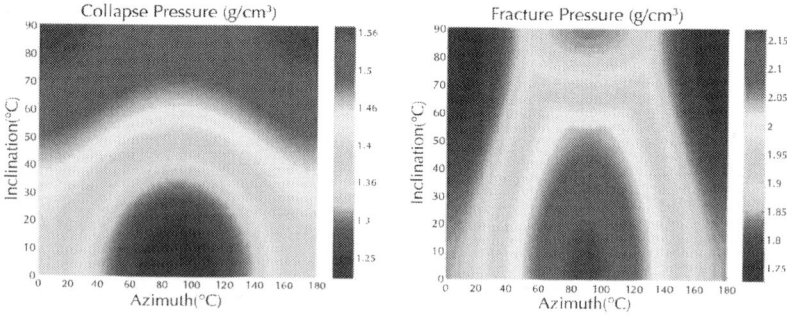

Figure 10: Effect of inclination angle and azimuth angle on critical mud weights window.

Thermal Effect on Horizontal Well

Fig. 11 shows the thermal effect on critical mud weight for vertical and horizontal wells. For vertical well, the breakdown mud weight and collapse mud weight change by 1.16×10^{-3} g/cm^3 and 5.81×10^{-4} g/cm^3 for every 1 °C variation in a vertical well, respectively. A linear relationship of thermal effect on critical mud weight window is also obtained for horizontal wells. However, the thermal effect on horizontal wells is more sensitive than on vertical wells. The breakdown mud weight decreases by 2.32×10^{-3} g/cm^3 for every 1 °C of decreasing for a horizontal well, while the collapse mud weight only has a 5.81×10^{-4} g/cm^3 decreasing, wellbore temperature variation has a greater effect on the formation breakdown pressure than on the formation collapse pressure for both vertical and horizontal wells.

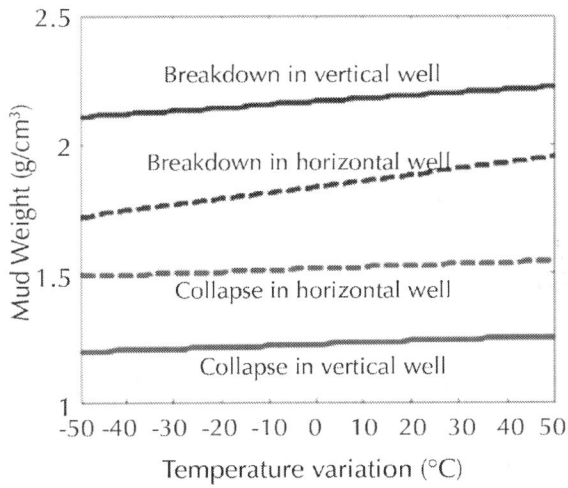

Figure 11: Effect of temperature on critical mud weight window for vertical and horizontal wellbores.

The wellbore is more apt to fracture when the formation temperature is decreased because the cooling effect will cause a tensile stress in circumferential direction which can reduce the hoop stress near wellbore. Increasing the formation temperature increases both the breakdown and collapse mud weights, however presents a smaller effect on collapse mud weight.

Combining with the true drilling condition, critical mud weights of horizontal well along the horizontal section with different mud system are shown in Fig. 12. For oil-based mud, the temperature at the toe of horizontal section is rise above the static formation temperature, and causes a larger critical mud weight window, while opposite situation occurs at the heel of horizontal section. When using the water-based mud, the temperature along the horizontal section of the wellbore is below the static formation temperature, which makes the critical mud weight window narrower and move downward, as shown in Fig. 12.

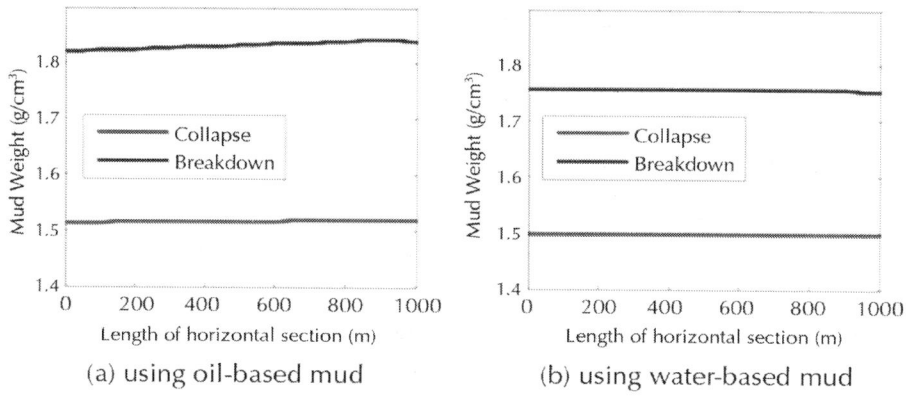

(a) using oil-based mud   (b) using water-based mud

Figure 12: Critical mud weights window of horizontal well along the horizontal section with different mud system.

Because of the sufficient heat exchange in the long horizontal section, the temperature variation magnitude in horizontal well is smaller than that in vertical by using either mud system, and the bottom hole temperature in horizontal well is more close to the static formation temperature. However, the effect of thermal stress on critical mud weight window in horizontal is more sensitive, therefore, the thermal effect on wellbore stability for different

24. Yoshioka, K., Zhu, D., Hill, A.D., Dawkrajai, P., Lake, L.W., 2007. Prediction of temperature changes caused by water or gas entry into a horizontal well. SPE Prod. Oper. 22 (4), 425e433.

25. Yu, M., Chen, G., Chenevert, M.E., 2001. Chemical and Thermal Effects on Wellbore Stability of Shale Formations. SPE 71366.

of such structures are usually invariant along their longitudinal direction; i.e. generalized plane strain. In such problems, material anisotropy and three-dimensional stress and strain state can be considered, but they have a functional dependence on only two spatial variables (Cheng, 1998). Therefore, only a two-dimensional domain discretization is needed for the numerical solutions. Bai et al. (1999) presented a generalized plane strain solution for dual-porosity poroelastic media. In this paper the naturally fractured formation is assumed as a dual-porosity and dual-permeability continuum medium, as shown in Fig. 1. The rock medium consists of equally spaced fractures separated by the matrix blocks. Fractures and matrix blocks have respective but distinctly different porosities and permeabilities.

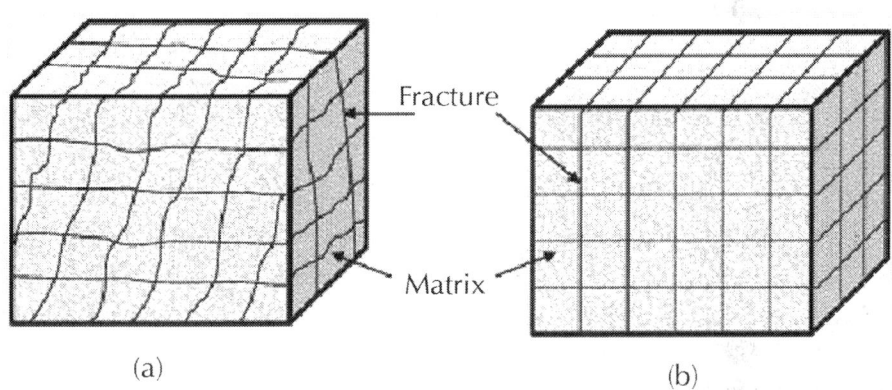

(a)   (b)

Figure 1: Sketch of the naturally fractured porous formation (a) and its simplification as a dual-porosity medium (b).

Formulations

For a separate and overlapping model, a double effective stress law needs to be considered. Combined with an elastic matrix and other related constants, the governing equations for solid (matrix and fractures) and fluid phases in the dual-porosity poromechanical formation can be obtained. In tensorial notation these equations are expressed as follows. For the solid phase the linear momentum conservation equation with fluid pressure effect is:

$$Gmfu_{i,jj}+(\lambda mf+Gmf)u_{k,ki}+\alpha ma DmfCma_{ijkl}pma_{,i}+\alpha frDmfCfr_{ijkl}pfr_{,i}=0.$$
(1a)

For the fluid phase in the matrix the diffusion equation with solid deformation effect is:

$$-\frac{k_{ma}}{\mu}p_{ma.kk} = \alpha_{ma}D_{mf}C_{maijkl}\frac{\partial \varepsilon_{kk}}{\partial t}-\beta_{ma}\frac{\partial p_{ma}}{\partial t}$$
$$+ \omega(p_{fr}-p_{ma}) + q_{ma}.$$
(1b)

For the fluid phase in the fractures the diffusion equation with solid deformation effect is:

$$-\frac{k_{fr}}{\mu}p_{fr.kk} = \alpha_{fr}D_{mf}C_{frijkl}\frac{\partial \varepsilon_{kk}}{\partial t}-\beta_{fr}\frac{\partial p_{fr}}{\partial t}$$
$$+ \omega(p_{fr}-p_{ma}) + q_{fr}$$
(1c)

where subscript 'ma' and 'fr' represent the matrix and fractures, respectively; i, j, k, l are the induces, and each takes the value of 1, 2, or 3; α is the Biot›s effective stress coefficient; p is the fluid pressure; k is the permeability; D_{mf} is the combined elasticity tensor; $C_{ma}ijkl$ and $C_{fr}ijkl$ are the compliance tensors for the rock matrix and the fracture systems, respectively; β is the relative compressibility representing the lumped deformability of the fluid and the solid; μ is the fluid dynamic viscosity; u is the solid displacement; ε_{kk} is the total body strain; q is the applied boundary flowrate; ω is the transfer coefficient, $\omega = 60k_{ma} / \mu s2$ for three mutually orthogonal fracture sets (Warren and Root, 1963); s is the fracture spacing; t is time; and λ_{mf} and G_{mf} are the Lamé's constants for the combined dual porosity media, which can be obtained from the following relationships:

$$G_{mf} = \frac{D_{mf}}{2(1 + v)}$$
(2)

$$\lambda_{mf} = \frac{vD_{mf}}{(1+v)(1-2v)} \tag{3}$$

where v is Poisson›s ratio.

The displacement in the solid phase can be solved from Eqs. (1a), (1b) and (1c). Then, according to the following strain–displacement and stress–strain relationships, the strains and stresses can be obtained:

$$\varepsilon_{ij} = (u_{i,j} + u_{j,i})/2 \tag{4}$$

$$\sigma_{ij} = D_{mf}\left(\varepsilon_{kl} + C_{maklmn}\alpha_{ma}p_{ma}\delta_{mn} + C_{frklmn}\alpha_{fr}p_{fr}\delta_{mn}\right) \tag{5}$$

where σ_{ij} is the total stress tensor; i, j are the induces and each takes the value of 1, 2, or 3; k, l, m, n are the induces and each takes the value of 1, 2, or 3. ε_{ij} and ε_{kl} are the strain tensors; and δ is the Kronecker delta.

The finite element method (FEM) proposed by Zhang (2002) and Zhang et al. (2003) is used to solve this coupled poromechanical formulation.

Fluid Mudweight Considerations at the Borehole Wall

In this finite element analysis, the borehole boundary conditions are obtained by subtracting the constant far-field stresses and pore pressures. By doing this on the outer boundary surfaces of the finite domain, all the tractions and pore pressure vanish, and at the borehole wall the equivalent stresses and pressures are given. After solving this problem, the final solution can be obtained by adding back the constant background stresses and pressures, as shown in Fig. 2 and Fig. 3.

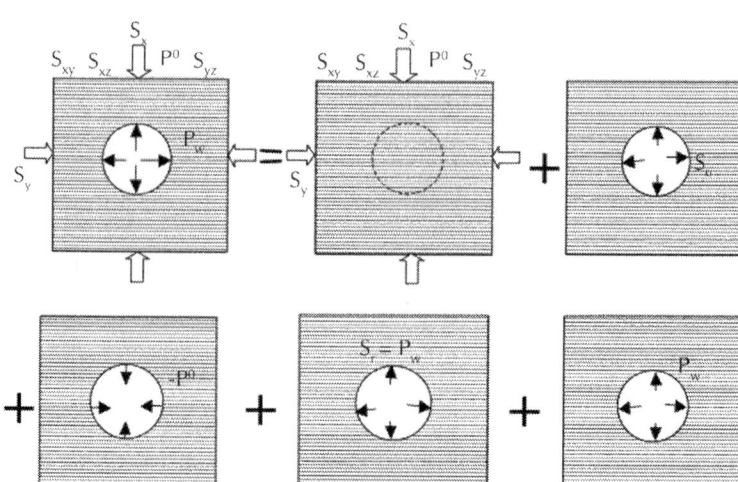

Figure 2: Superposition of mudweight pressure at the borehole wall for the permeable case.

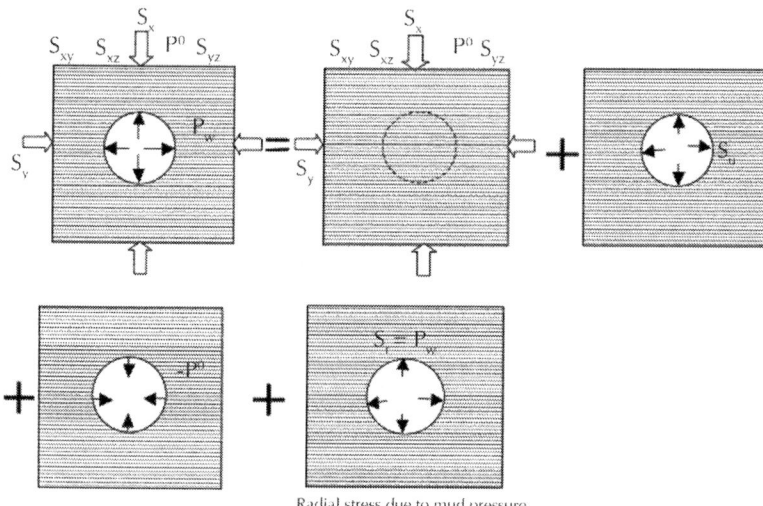

Figure 3: Superposition of mudweight pressure at the borehole wall for the impermeable case.

Permeable Boundary

During drilling, the suitable mudweight of the drilling fluid plays a very important role in protecting the borehole wall from breakouts and failures. The mudweight pressure, p_w, acts both as a radial stress (S_r) and a fluid pressure at the borehole wall when the wellbore formation is permeable. Fig. 2 illustrates the superposition principle allowing one to consider the mudweight pressure at the permeable borehole wall for the dual-porosity finite element modeling, In this figure p^0 represents both the initial pore and fracture pressures. Considering mud pressure effects, the equivalent stresses (σ_x, σ_y, τ_{xy}) and pressures at any node at the borehole wall can be expressed as (Zhang et al., 2003):

$$\sigma_x = l(S_x - S_{mx}) + m(S_{yx} - S_{myx}) + nS_{zx} \tag{6}$$

$$\sigma_y = m(S_y - S_{my}) + nS_{zy} + l(S_{xy} - S_{mxy})$$

$$\tau_{xy} = nS_z + lS_{xz} + mS_{yz}$$

$$p_{ma} = -p^0_{ma} + p_w$$

$$p_{fr} = -p^0_{fr} + p_w$$

where *l*, *m*, and *n* are the direction cosines between the normal to the inclined plane and the x-, y-, andz-axes, respectively; S_x, S_y, S_z, S_{xy}, S_{yz}, and S_{zx} are the stresses in the local borehole (FEM mesh) coordinate; p_{ma} and p_{fr} are the fluid pressures in the matrix and fractures, respectively; p_{ma}^0 and p_{fr}^0 are the fluid pressures at the borehole wall for the matrix and fractures, respectively; S_{mx}, S_{my}, and S_{myx} are stresses induced by the mud pressure (p_w) acted as the radial stress, which can be obtained by:

$$S_{m\boxtimes x} = l^2 p w \quad S_{m\boxtimes y} = m^2 p w \quad S_{m\boxtimes x\boxtimes y} = l\boxtimes m p w. \tag{7}$$

Impermeable Boundary

A mud cake can be formed during drilling which may make the borehole wall impermeable. In this case, the flowrate at the borehole wall is zero and the mudweight pressure, p_w, acts only as a radial stress at the borehole wall (Fig. 3). Therefore, the equivalent stresses and pressures at any node at the borehole wall can be written as:

$$\sigma_x = l(S_x - S_{mx}) + m(S_{yx} - S_{myx}) + nS_{zx}$$

$$\sigma_y = m(S_y - S_{my}) + nS_{zy} + l(S_{xy} - S_{mxy})$$

$$\tau_{xy} = nS_z + lS_{xz} + mS_{yz}$$

$$p_{ma} = -p_{ma}$$

$$p_{fr} = -p_{fr}^0 \qquad (8)$$

where S_{mx}, S_{my}, and S_{myx} can be obtained by Eq. (7).

ROCK FAILURE CRITERIA

The Biot's effective stress coefficient is considered to calculate effective stresses in this paper. However, it is well-known that rock failure is controlled by the Terzaghi's effective stresses, because Biot's coefficient approaches 1.0 when rock failure is approached. Therefore in the failure criteria the Terzaghi's effective stresses (refer to Eq. (9)) need to be used.

$$\sigma_{ij}' = \sigma_{ij} - \delta_{ij}p_{ma}$$

$$\qquad (9)$$

where σ_{ij}' is the effective stress tensor.

Depending upon the in-situ stress environments and combinations of their relative magnitudes, various failure criteria may be required at the specified field conditions to delineate most wellbore stability situations. Therefore, some typical related failure criteria are briefly introduced.

Mohr–coulomb Failure Criterion

In the principal space $(\sigma_1', \sigma_2', \sigma_3')$, it can be expressed as:

$$\sigma_1' = \sigma c + q \sigma_3', \tag{10}$$

where σ_1', σ_3' are the effective maximum and minimum principal stresses, respectively; σ_c is the rock uniaxial compressive strength, $q = (1 + \sin\varphi) / (1 - \sin\varphi)$; and φ is the angle of internal friction.

The effective Mohr–Coulomb failure stress can thus be defined as:

$$\sigma mohr = \sigma c + q \sigma_3' - \sigma_1', \tag{11}$$

where σ_{mohr} is the effective Mohr–Coulomb failure stress. When this value becomes negative, the rock fails.

Drüker–Prager Failure Criterion

Laboratory data have shown that the intermediate principal stress can play an important stabilization role in the failure of rocks. This can be conveniently represented by the Drüker–Prager yield condition, sometimes referred to as the collapse criterion:

$$\sqrt{J_2} = 3\alpha I_1' + \kappa, \tag{12}$$

where α and κ are material constants; J_2 is the stress invariant of the stress deviator; I_1' is the first effective stress invariant; and,

$$J_2 = \frac{1}{6}\left[(\sigma_x-\sigma_y)^2 + (\sigma_y-\sigma_z)^2 + (\sigma_z-\sigma_x)^2\right] + \tau_{xy}^2$$
$$+\tau_{yz}^2 + \tau_{zx}^2,$$

$I_1' = \sigma x' + \sigma y' + \sigma z',$

where σ_x', σ_y' and σ_z' are the effective normal stresses; τ_{xy}, τ_{yz}, and τ_{zx} are the shear stresses.

The effective collapse stress, σ_{dp} can be defined as:

$$\sigma_{dp} = 3\alpha I_1' + \kappa - \sqrt{J_2},$$

(13)

when the effective collapse stress is negative, the rock is subjected to collapse failure.

HORIZONTAL WELL STABILITY IN A STRIKE SLIP FAULTING STRESS REGIME

The studied area was located in the western Canadian over-thrust belt, where the vertical stress is not the maximum principal stress. The in-situ stresses and initial pore pressures were (Woodland, 1990): σ_H = 29 MPa, σ_h = 20 MPa, σ_v = 25 MPa, and the far field pore and fracture pressures $p^0 = p_{ma}{}^0 = p_{fr}{}^0$ = 10 MPa. In the local coordinate system (i.e. the FEM mesh), the stresses and fluid pressures are presented in Fig. 4 for the horizontal wells drilled along the maximum and minimum directions, respectively. The wellbore radius was R = 0.1 m and the loads on the wellbore are assumed to be applied instantaneously. The formation is assumed to be isotropic, linear elastic and characterized by the following poroelastic properties (Zhang, 2002 and Zhang and Roegiers, 2002): Biot modulus, M = 15.8 GPa; Biot's effective stress coefficients, α_{ma} =

0.771 and α_{fr} = 0.91; and fluid dynamic viscosity, μ = 0.001 Pa·s. The corresponding equivalent parameters for the dual-porosity poroelastic model are listed in Table 1.

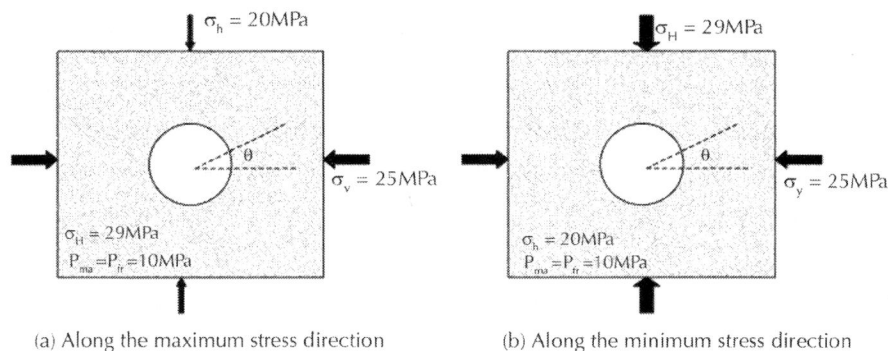

(a) Along the maximum stress direction   (b) Along the minimum stress direction

Figure 4: States of stress in the local coordinate system for horizontal boreholes drilled in two different directions.

Table 1: Parameters for boreholes in tectonic stress regimes

Parameter	Unit	Magnitude
Elastic modulus (E)	GN/m²	20.6
Poisson's ratio (v)	–	0.189
Fracture stiffness (K_n, K_{sh})	MN/m²/m	4.821×10^5
Fluid bulk modulus (K_f)	MN/m²	419.17
Grain bulk modulus (K_s)	GN/m²	48.21
Matrix porosity (n_{ma})	–	0.02
Fracture porosity (n_{fr})	–	0.002
Matrix mobility (k_{ma}/μ)	M⁴/MN s	10^{-10}
Fracture mobility (k_{fr}/μ)	M⁴/MN s	10^{-9}
Fracture spacing (s)	m	1
Internal friction angle (φ)	°	30
Tensile strength	MN/m²	1.5
Material constants (α)	–	0.14
Material constants (κ)	MN/m²	12

Dual-porosity Effects

The single- and dual-porosity poroelastic effects are illustrated by comparing the radial stresses around the wellbore for a horizontal well drilled along the maximum stress direction. The Terzaghi's effective radial stress distributions at $t = 100$ s and $\theta = 90°$ (refer to Fig. 4a) for the elastic and single-porosity models, as well as the one for the dual-porosity model, are presented in Fig. 5. It can be observed that for both single- and dual-porosity models without mudweight support, tensile stresses are induced at a small distance inside the borehole wall, evidence of poroelastic effects. This tends to cause wellbore spalling because the rock has a very low tensile strength.

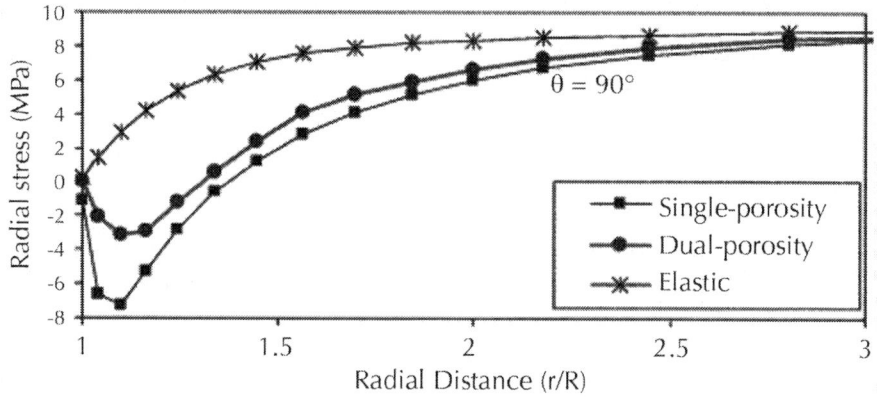

Figure 5: Comparison of effective radial stress for elastic, single- and dual-porosity models for the horizontal hole drilled in the maximum stress direction.

The decrease of the near-well tensile stress in the dual-porosity model (Fig. 5) is due to the fact that the total radial stress has a larger increment. Therefore, the effective compressive stress increases and the effective tensile stress decreases, which reduces the potential for borehole spalling. The reason for this phenomenon is that the total deformation increases due to the introduction of the combined elastic modulus, D_{mf} and fracture compliance, C_{fr} in the dual-porosity governing equation (refer to Eq. (5)), which leads to an increase in the total radial stress.

Influence of Borehole Drilling Directions

Two different drilling directions are considered in the ongoing analysis: in the maximum and minimum horizontal stress directions. Fig. 6 gives the effective tangential stresses. It can be seen that, when the borehole axis is along the maximum stress direction, the tangential stress is much smaller. It implies that it is much easier to induce compressive failure for the horizontal borehole drilled along the minimum stress direction.

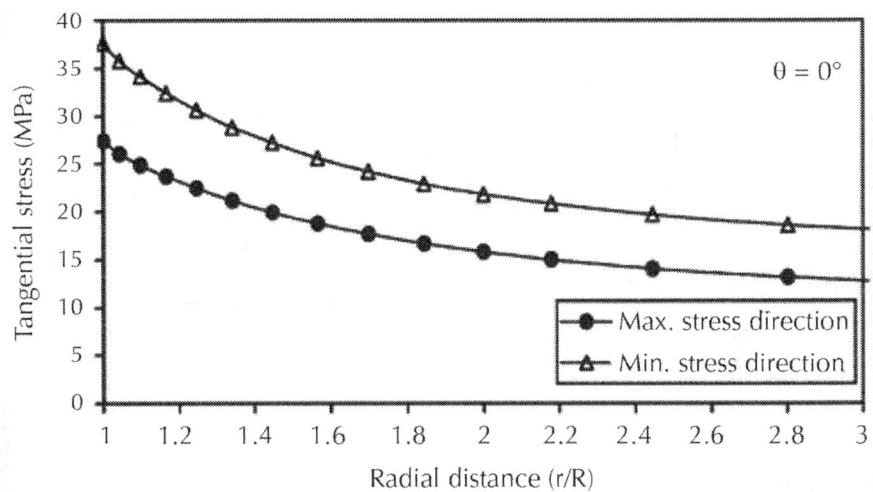

Figure 6: Effective tangential stresses at the borehole section of $\theta = 0°$ for two different borehole drilling directions.

Fig. 7 plots the effective radial stresses (spalling stresses) for the permeable wellbore boundary at $\theta = 0°$. It can be seen that effective radial stresses have smaller magnitudes when the borehole axis is along the maximum stress direction: hence less spalling failure probability.

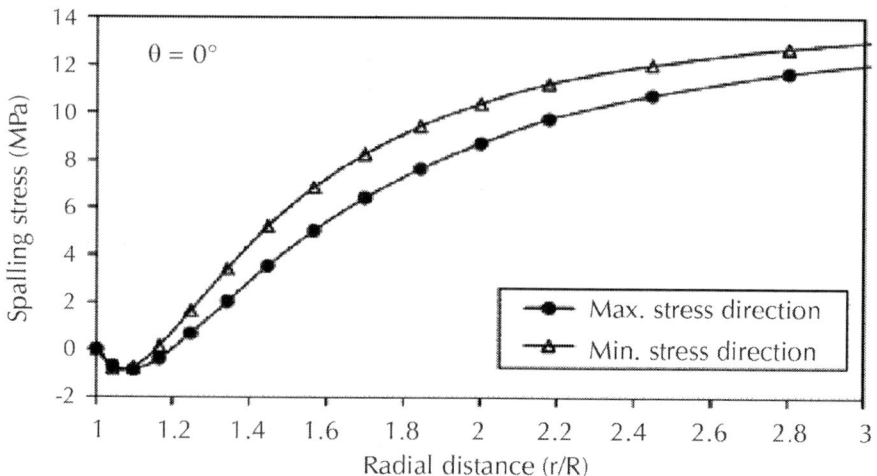

Figure 7: Effective radial stresses at the borehole section of θ = 0° for two different borehole drilling directions.

Fig. 8 shows effective collapse stresses at θ = 0° for horizontal wells without mudweight support. It can be seen that the collapse stress has large negative values (meaning failure) for a borehole drilled along the minimum horizontal stress direction. It may be noted that there is no collapse failure in the other case. Fig. 9exhibits similar trends at θ = 90°. These indicate that it is much easier to collapse a horizontal well drilled along the minimum horizontal stress direction.

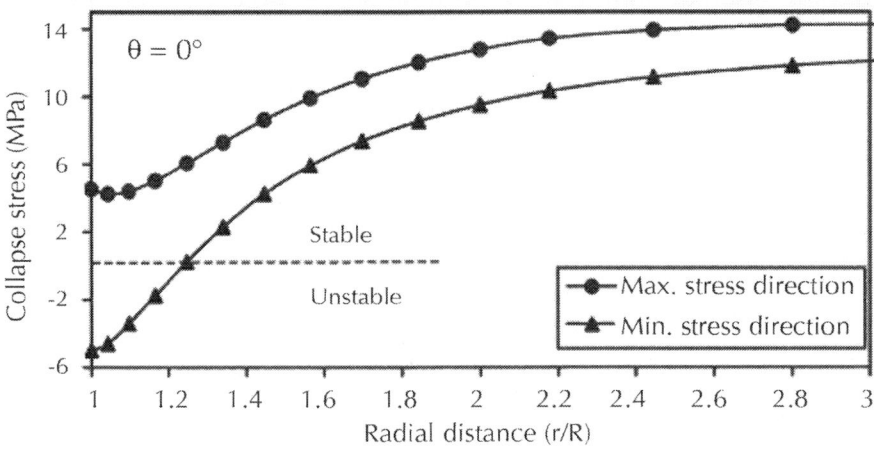

Figure 8: Effective collapse stresses at the borehole section of = 0° for two different borehole drilling directions.

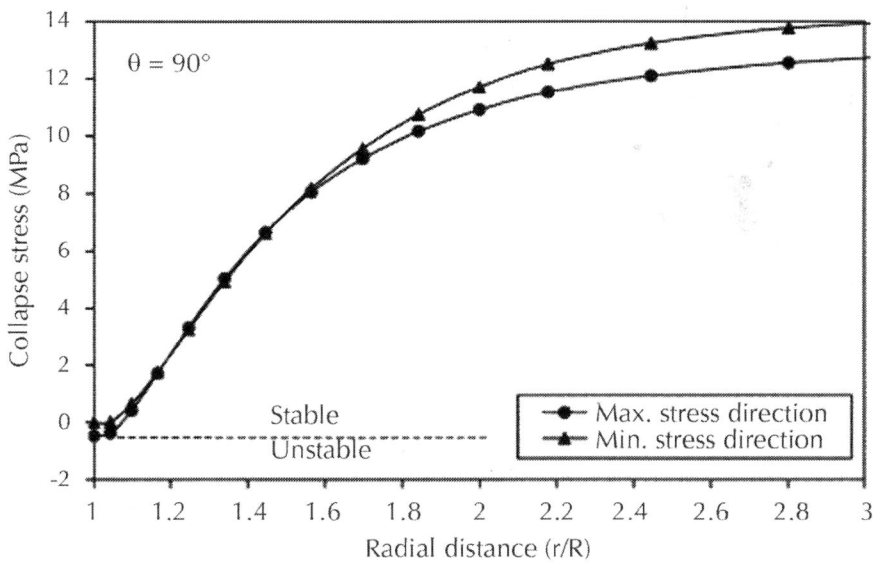

Figure 9: Effective collapse stresses at the borehole section of = 90° for two different borehole drilling directions.

The collapsed (i.e. Drüker–Prager failure) area is given in Fig. 10 for drilling along the maximum horizontal stress direction without mudweight support. It can be seen that the failure zone is concentrated around θ = 90° of the borehole section (breakout angle of about 78°). However, for the borehole drilled parallel to the minimum horizontal stress, the failure zone (as shown in Fig. 11) is distributed around all the wellbore over a much larger area. Comparing the local stress configurations in the far-field in the cross section (Fig. 4), the stress difference for the borehole in the maximum horizontal stress direction is $\Delta\sigma$ = 5 MPa (Fig. 4a), while the stress difference in the minimum stress direction is only $\Delta\sigma$ = 4 MPa (Fig. 4b); but, the latter results in a much larger failure area! Therefore, a larger stress difference in the far-field does not necessarily induce larger borehole failures, which is not consistent with the conventional understanding.

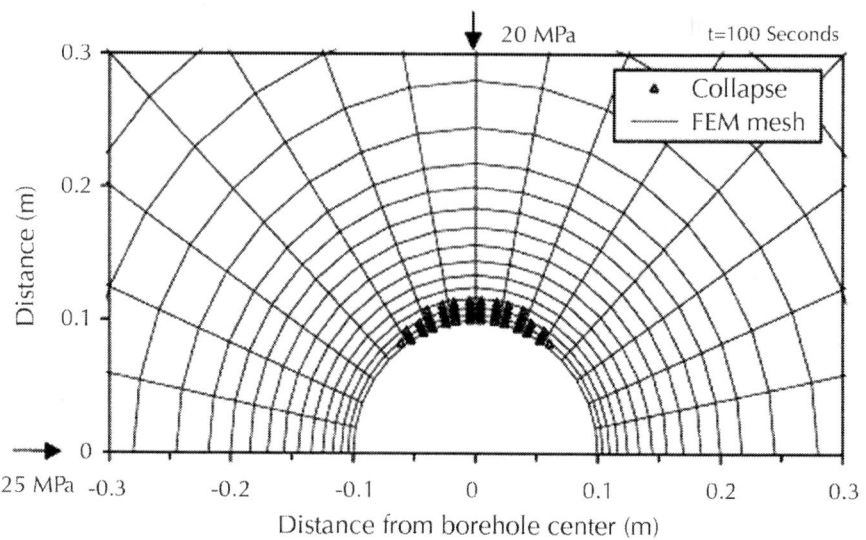

Figure 10: Collapsed area in the strike slip faulting stress regime at t = 100 s for a horizontal borehole drilled along the maximum horizontal stress direction.

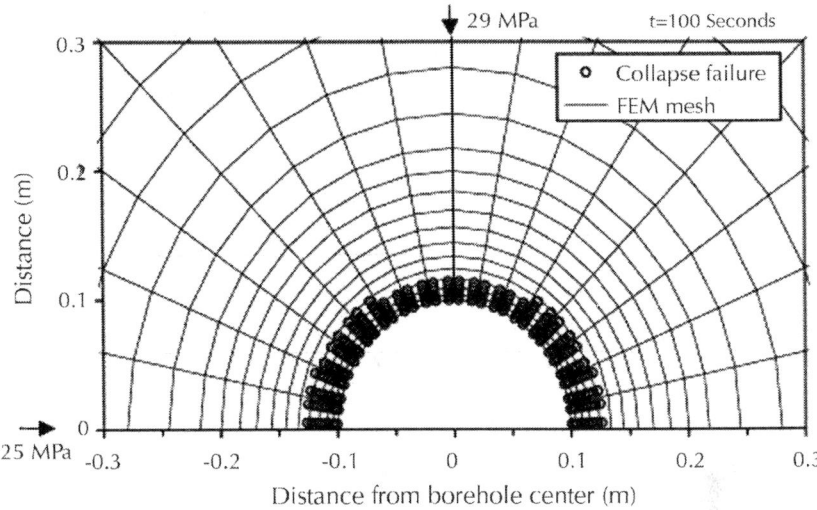

Figure 11: Collapsed area in the strike slip faulting stress regime at $t = 100$ s for a horizontal borehole drilled along the minimum horizontal stress direction.

Mud Pressure Effects

The instability due to collapse and radial tensile stresses (or spalling stresses) can be avoided by selecting proper mudweights. Fig. 12 represents these stresses at a section of $\vartheta = 90°$ for the case shown in Fig. 10, but with mud pressure support ($p_w = 12$ MPa). It is obvious that mud prevents instabilities.

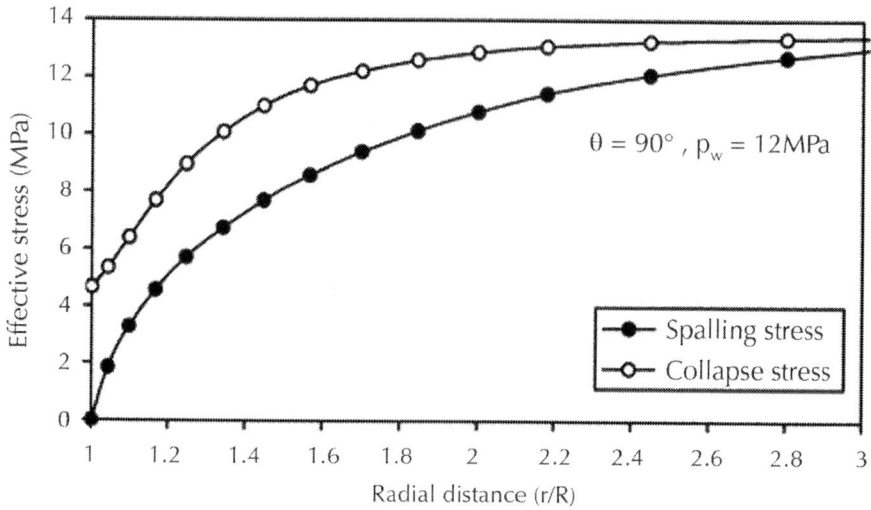

Figure 12: Collapse and spalling stresses at the borehole section of = 90° for horizontal hole drilled in the maximum horizontal direction with mudweight.

Stability during Production

The selected mud pressure can keep the borehole from failing during drilling. However, most horizontal wells have open-hole completions; therefore, open-hole stability is of critical importance for safe and economic production. During open-hole production the bottomhole pressure will equal the pore/fluid pressure. Fig. 13 represents the collapse area for a horizontal well drilled in the minimum stress direction for a given gravity and reservoir conditions. It is observed that during open-hole production the horizontal well drilled in the minimum stress direction (Fig. 13) is unstable. However, from Fig. 14 a horizontal well drilled along the maximum stress direction keeps stable during open-hole production (with the bottomhole pressure of 10 MPa). Similar conclusions were obtained for the Mohr–Coulomb failure criterion (Zhang, 2002).

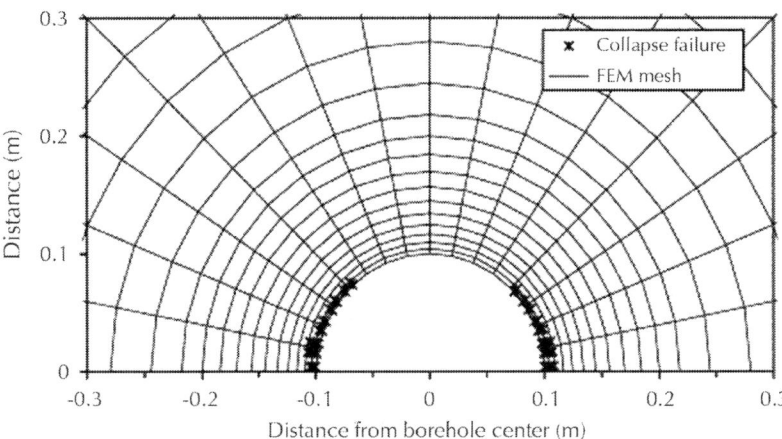

Figure 13: Collapsed area in the strip slip faulting stress regime for a horizontal borehole drilled parallel to the minimum horizontal stress during production (bottomhole pressure = 10 MPa).

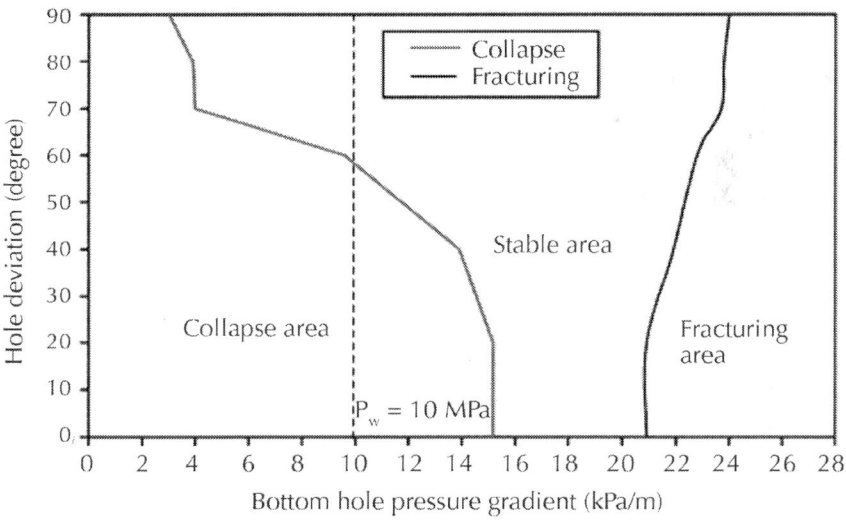

Figure 14: Bottomhole pressure range varying with hole inclinations for collapse and fracturing at $t = 100$ s in a strike slip faulting stress regime (in this figure 90° corresponds to a horizontal well).

Therefore, it can be concluded that in a strike slip faulting stress regime a horizontal borehole drilled along the maximum stress direction is probably, in most cases, more stable during both drilling and production stages, which is contrary to the traditional drilling guidelines. Field studies verify this conclusion, for example, actual drilling experience in a field in a sub-Andean foreland basin in northwestern South America showed that the most stable drilling direction was parallel to the maximum horizontal stress with a high deviation (near-horizontal) (Zoback et al., 2003).

HORIZONTAL WELL STABILITY IN A NORMAL FAULTING STRESS REGIME

The following in-situ stress configurations are used for analyzing horizontal well stability in a normal faulting stress regime; i.e. $\sigma v = 69$ MPa, $\sigma_H = 55.2$ MPa, $\sigma_h = 48.3$ MPa, and the pore and fracture fluid pressures $p^0 = p_{ma}^0 = p_{fr}^0 = 31.7$ MPa (Zheng, 1998). The material parameters are listed in Table 2. In order to demonstrate the stability characteristics of horizontal wells of different orientations, shear and collapse failures in a stress-free wellbore (no support) are first examined. The stability of horizontal wells during open-hole productions is then analyzed.

Table 2: Parameters for boreholes in a normal stress regime

Parameter	Unit	Magnitude
Elastic modulus (E)	GN/m^2	18.53
Poisson's ratio (v)	–	0.219
Fracture stiffness (K_n, K_{sh})	MN/m^2/m	4.821 × 10^5
Fluid bulk modulus (K_f)	MN/m^2	173.45
Grain bulk modulus (K_s)	GN/m^2	323.23
Matrix porosity (n_{ma})	–	0.02

Fracture porosity (n_{fr})	–	0.002
Matrix mobility (k_{ma}/μ)	M⁴/MN s	10^{-10}
Fracture mobility (k_{f}/μ)	M⁴/MN s	10^{-9}
Fracture spacing (s)	m	1
Tensile strength	MN/m²	1.5
Uniaxial compressive strength (σ_c)	MN/m²	41
Internal friction angle (φ)	°	30
Material strength parameter (α)	–	0.14
Material strength parameter (κ)	MN/m²	18

Mohr–Coulomb Failure

Fig. 15 and Fig. 16 are the Mohr–Coulomb failure areas for horizontal boreholes drilled along two directions without mudweight support. Comparing the two figures indicates that there is a slightly larger failure area for the hole drilled in the maximum horizontal direction. This is due to the fact that in the Mohr–Coulomb failure criterion only the maximum and minimum principal stresses are considered, and drilling in the minimum stress direction reduces the local far-field stress difference. In Fig. 17, it is clear that the effective Mohr–Coulomb stress has a larger negative magnitude, meaning less stable, for a hole drilled along the maximum direction. Fig. 18 and Fig. 19 show the failure areas in both those directions during open-hole production, in which the bottomhole pressure is 31.7 MPa. It can be seen that there is a slightly larger failure for a hole drilled along the minimum horizontal stress direction. Thus, it can be concluded that, in a normal stress regime, the hole drilled along the minimum stress direction has been subjected to slightly less Mohr–Coulomb failure during drilling, but, to a slightly larger failure zone during open-hole production.

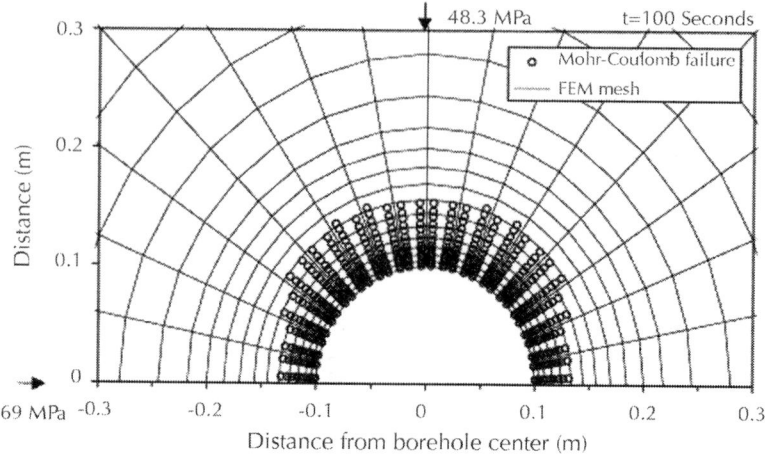

Figure 15: Mohr–Coulomb failure area in a normal faulting stress regime at t = 100 s for a horizontal borehole drilled along the maximum horizontal stress direction.

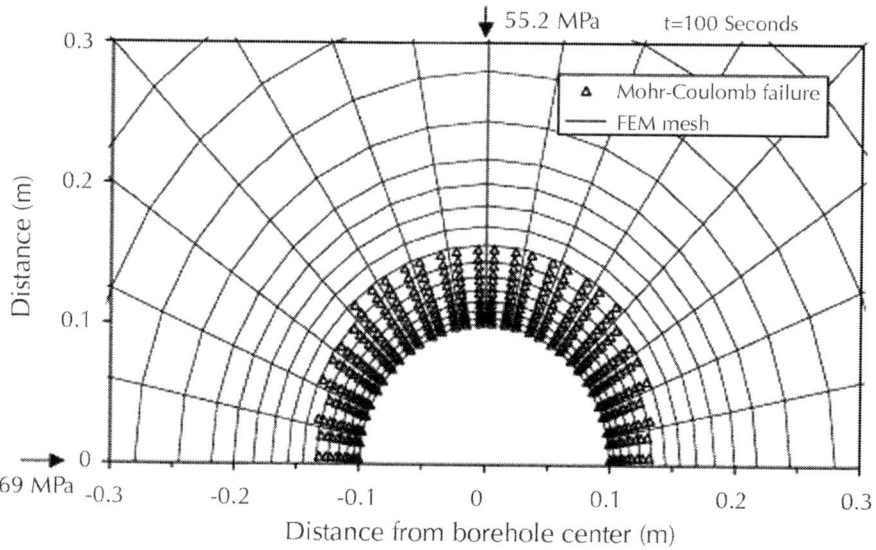

Figure 16: Mohr–Coulomb failure area in a normal faulting stress regime at t = 100 s for a horizontal borehole drilled along the minimum horizontal stress direction.

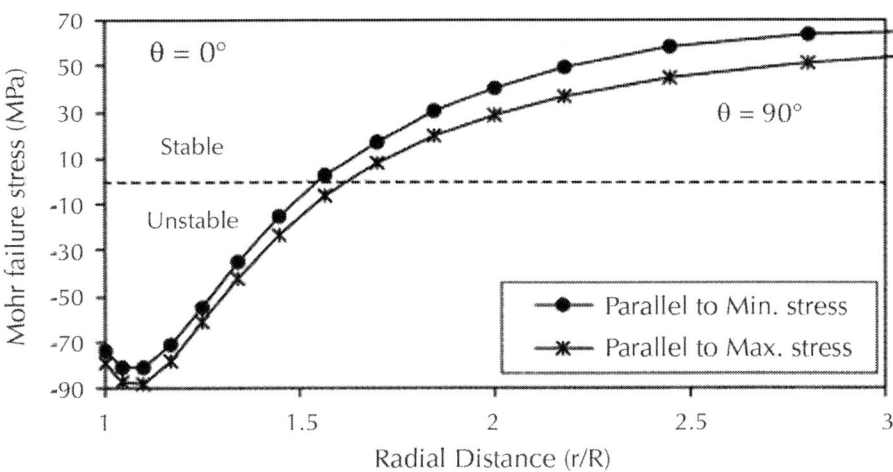

Figure 17: Mohr–Coulomb failure stresses at the hole section of = 90° at *t* = 100 s for a borehole without support, drilled along the minimum and maximum horizontal stress directions.

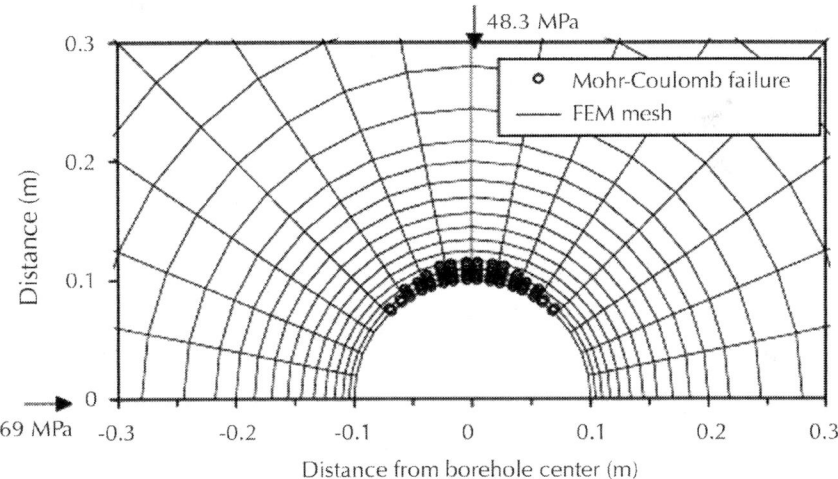

Figure 18: Mohr–Coulomb failure area in a normal faulting stress regime at *t* = 100 s for a horizontal borehole drilled along the maximum horizontal stress direction during open-hole production (bottomhole pressure 31.7 MPa).

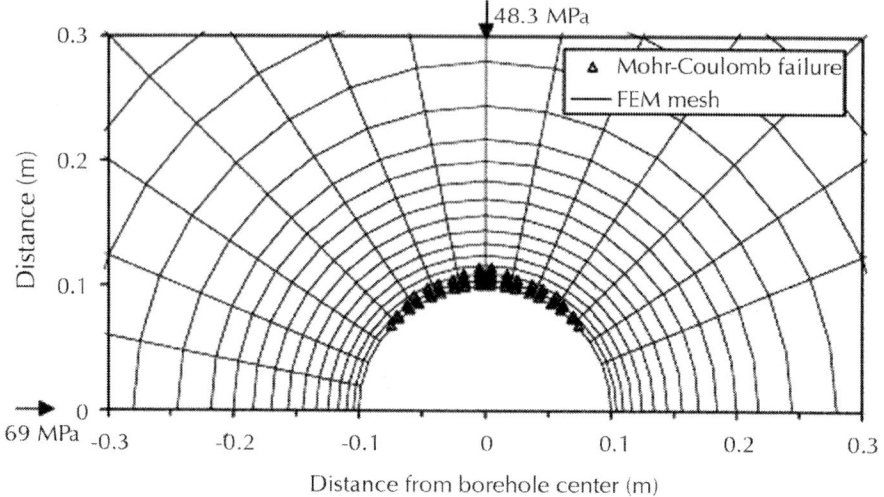

Figure 19: Mohr–Coulomb failure area in a normal faulting stress regime at $t = 100$ s for a horizontal borehole drilled along the minimum horizontal stress direction during open-hole production (bottomhole pressure 31.7 MPa).

Drüker–Prager Failure

The collapsed (Drüker–Prager failure) areas are given in Fig. 20 and Fig. 21 during the drilling stage without mudweight support. Fig. 22 and Fig. 23 are the failure areas during open-hole production with a bottomhole pressure of 31.7 MPa. It can be observed that in both drilling and open-hole production stages the collapse areas are always larger for the hole drilled along the minimum horizontal stress direction (refer to Fig. 20, Fig. 21, Fig. 22 and Fig. 23).

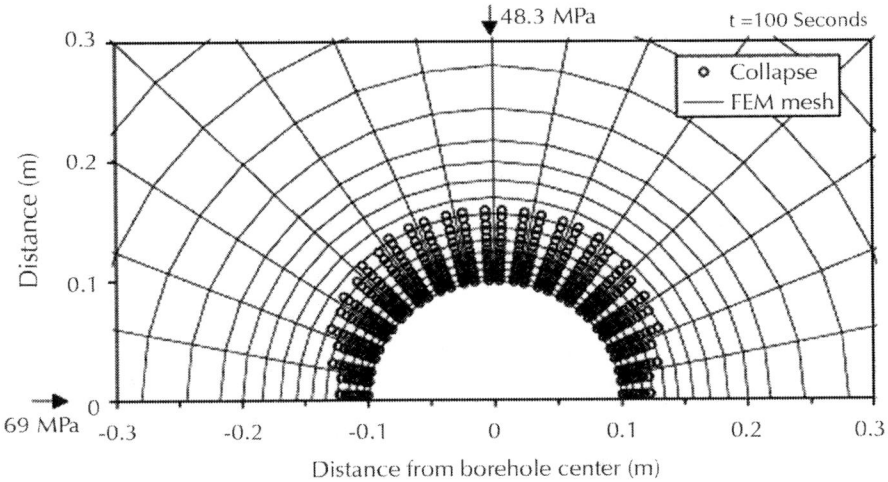

Figure 20: The collapse in a normal faulting stress regime at $t = 100$ s for a horizontal borehole drilled along the maximum horizontal stress direction.

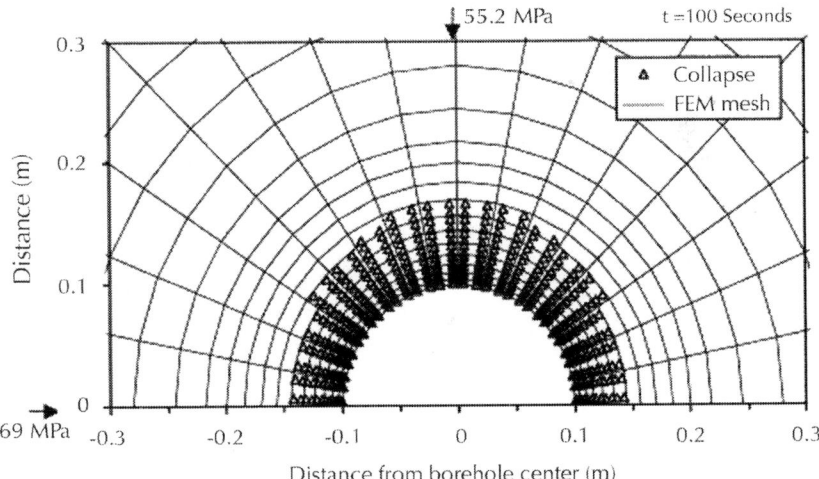

Figure 21: The collapse in a normal faulting stress regime at $t = 100$ s for a horizontal borehole drilled along the minimum horizontal stress direction.

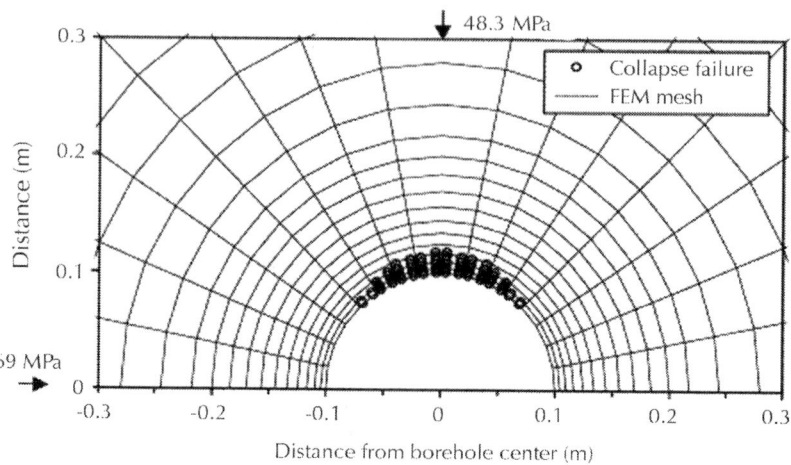

Figure 22: The collapse area in a normal faulting stress regime at $t = 100$ s for a horizontal borehole drilled along the maximum horizontal stress direction during open-hole production (bottomhole pressure 31.7 MPa).

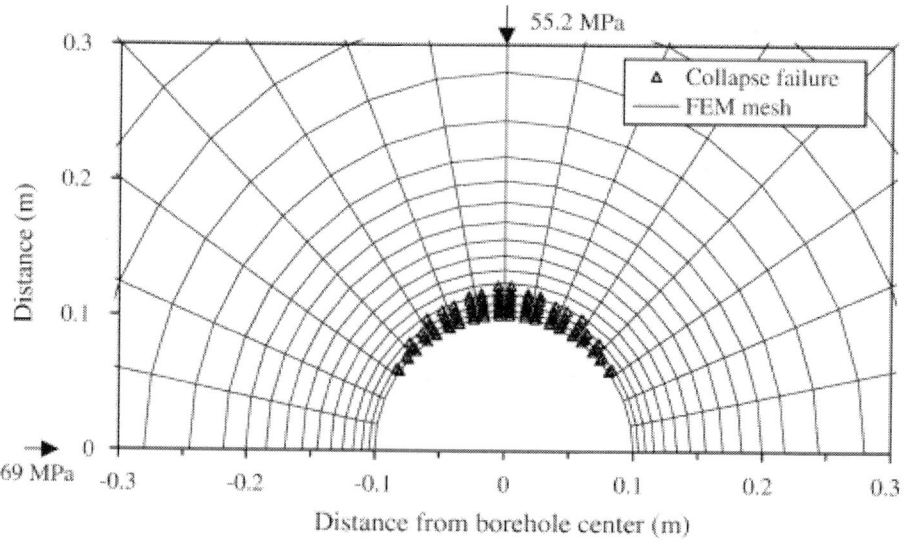

Figure 23: The collapse area in a normal faulting stress regime at $t = 100$ s for a horizontal borehole drilled along the minimum horizontal stress direction during open-hole production (bottomhole pressure 31.7 MPa).

According to this study, the collapsed areas during the drilling stage are plotted for the wells drilled in the maximum and minimum horizontal stress directions (Fig. 24 and Fig. 25). It can be seen that the stress configuration differences produced by different drilling directions and in-situ state of stress have a significant effect on the wellbore failure and instability.

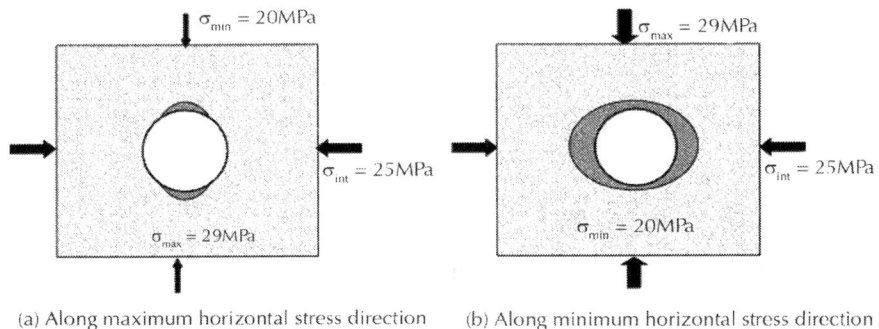

(a) Along maximum horizontal stress direction   (b) Along minimum horizontal stress direction

Figure 24: Schematic collapsed zone in a strike slip faulting stress regime for horizontal wells drilled in different horizontal stress directions.

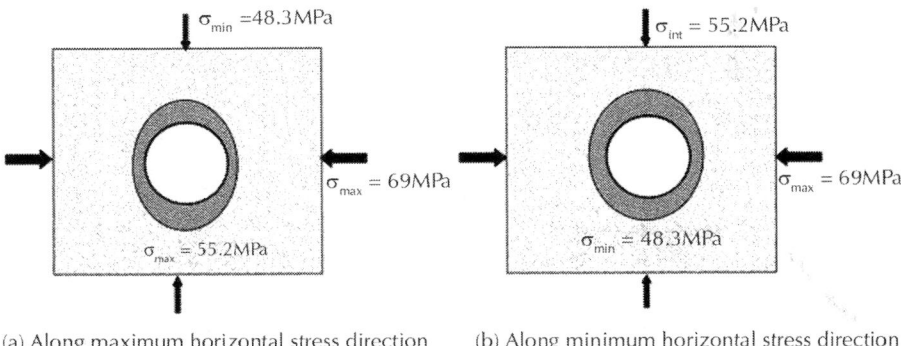

(a) Along maximum horizontal stress direction   (b) Along minimum horizontal stress direction

Figure 25: Schematic collapsed zone in a normal faulting stress regime for horizontal wells drilled in different horizontal stress directions.

CONCLUSIONS

A dual-porosity finite element model is presented and applied to analyze horizontal well stability in naturally fractured reservoirs. Particularly, when the wells are located in the active tectonic regimes, the pay zones can be dominated by fractures. In this case, the dual-porosity poroelastic model is suitable to describe wellbore mechanical and hydraulic behaviors.

Two numerical studies are performed for horizontal wells drilled in normal faulting and strike slip faulting stress regimes with different drilling orientations. It is found that stress concentrations and well stability in horizontal wells depend strongly upon the in-situ stress regime and drilling directions.

The studies have shown that in a strike slip faulting stress regime, when the borehole longitudinal axis direction is coincident with the maximum horizontal stress direction, large collapse and spalling failures can be avoided. That is, in a strike slip faulting regime a horizontal borehole drilled along the maximum stress direction is more stable during both drilling and open-hole production stages.

During the drilling in the normal faulting stress regime, however, the horizontal well may have a slightly larger Mohr–Coulomb failure zone and a slightly smaller collapse zone for the well drilled in the maximum horizontal stress direction. This apparent contradiction stems from the fact that the intermediate principal stress is considered in the Drücker–Prager failure criterion.

It is important for the well to be kept stable during the production stage, because suitable mud weight pressure can prevent horizontal well instabilities during drilling. For the cases with two different horizontal stresses, the well is more stable during open-hole production when the borehole longitudinal axis is in the maximum horizontal stress direction.

ACKNOWLEDGMENTS

Authors would like to thank Dr. Matt Matthews and Mr. William Standifird at Knowledge Systems, Inc. for their reviews and constructive suggestions to improve the manuscript. Suggestions and comments from editors and anonymous reviewers are gratefully acknowledged.

REFERENCES

1. Bai, M., Abousleiman, Y., Cui, L., Zhang, J., 1999. Dual-porosity poroelastic modeling of generalized plane strain. Int. J. Rock Mech. Min. Sci. 36 (8), 1087–1094.

2. Bai, M., Elsworth, D., Roegiers, J.-C., 1993. Multi-porosity/multipermeability approach to the simulation of naturally fractured reservoirs. Water Resour. Res. 29 (6), 1621–1633.

3. Bai, M., Roegiers, J.-C., Elsworth, D., 1995. Poromechanical response of fractured-porous rock masses. J. Pet. Sci. Eng. 13, 155–168.

4. Barenblatt, G.I., Zheltov, I.P., Kochina, N., 1960. Basic concepts in the theory of seepage of homogeneous liquids in fissured rocks. Prikl. Mat. Mekh. 24 (5), 852–864.

5. Bradley,W.B., 1979. Failure of inclined boreholes. Trans. ASME 101, 232–239.

6. Chen, G., Chenevert, M.E., Sharma, M.M., Yu, M., 2003a. A study of wellbore stability in shales including poroelastic, chemical, and thermal effects. J. Pet. Sci. Eng. 38, 167–176.

7. Chen, X., Tan, C.P., Detournay, C., 2003b. A study on wellbore stability in fractured rock masses with impact of mud infiltration. J. Pet. Sci. Eng. 38, 145–154.

8. Cheng, A.H.-D., 1998. On generalized plane strain poroelasticity. Int. J. Rock Mech. Min. Sci. 35, 183–193.

9. Elsworth, D., Bai, M., 1992. Flow-deformation response of

dualporosity media. J. Geotech. Eng. 118 (1), 107–124.

10. Fung, L.S.K.,Wan, R.G., Rordriguez, H., Silva-Bellorin, R., Zerpa, L., 1999. Advanced elasto-plastic model for borehole stability analysis of horizontal wells in unconsolidated formation. J. Can. Pet. Technol. 38 (12), 41–48.

11. National Research Council, Stability, 1993. Failure, and Measurements of Boreholes and Other Circular Openings. National Academy Press, Washington, D.C.

12. Pruess, K., Tsang, Y.W., 1990. On two-phase relative permeability and capillary pressure of rough-walled rock fractures. Water Resour. Res. 26 (9), 1915–1926.

13. Tan, C.P., Detournay, C., Chen, X., 2005. Factors governing mud infiltration and impact on wellbore stability in fractured rock mass. Paper ARMA/USRMS05-834 presented at 40th U.S. Symp. Rock Mechanics, Anchorage, Alaska, USA.

14. Wang, Y., Dusseault, M.B., 2003. A coupled conductive-convective thermo-poroelastic solution and implications for wellbore stability. J. Pet. Sci. Eng. 38, 187–198.

15. Warren, J.E., Root, P.J., 1963. The behavior of naturally fractured reservoirs. SPEJ 228, 244–255.

16. Woodland, D.C., 1990. Borehole instability in the western Canadian overthrust belt. SPE Drill. Eng. 5, 23–33.

17. Zhang, J., 2002. Dual-porosity approach to wellbore stability in naturally fractured reservoirs. PhD Dissertation, Univ. of Oklahoma.

18. Zhang, J., Roegiers, J.-C., 2002. Borehole stability in naturally fractured reservoirs—a fully coupled approach. Paper SPE 77355 presented at the SPE Annual Technical Conf. Exhib. held in San Antonio.

19. Zhang, J., Bai, M., Roegiers, J.-C., 2003. Dual-porosity poroelastic analyses of wellbore stability. Int. J. Rock Mech. Min. Sci. 40 (4), 473–483.

20. Zheng, Z., 1998. Integrated borehole stability analysis—against tradition. SPE 47282 presented at SPE/ISRM Eurock'98, Trondheim, Norway.

21.   Zoback, M.D., et al., 2003. Determination of stress orientation and magnitude in deep wells. Int. J. Rock Mech. Min. Sci. 40, 1049–1076.

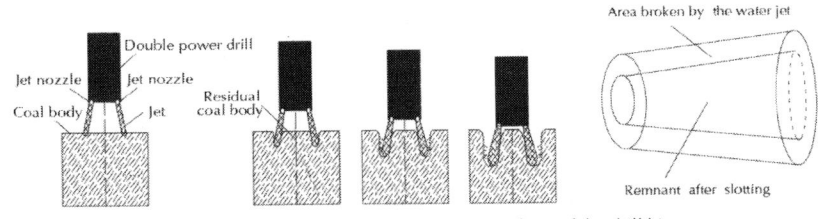

Figure 1: Double power coal breaking: the advance of the broken zone.

The individual processes are as follows. First, before the drill contacts the coal the high speed jet affects it. The coal breaks and erodes in front of the drill bit. Rotation of the drill forms a circular broken ring in conjunction with the slotting jet. As the drill advances the target distance of jet decreases and its breaking ability become stronger. The depth of the broken coal increases and so does the quantity of broken coal. Because the preceding jet becomes submerged its breaking ability is reduced but its erosion effect increases. As the jet advances there is a "breakup region" as shown in Fig. 1b. This is a bell shaped annular gap in front of the drill bit.

Second, as drilling continues a mix of breaking processes occurs. The bit begins to break the coal left from step 1 and the jet begins to affect the gap and expand it. This will help the bit break the residual coal that is then promptly taken away by the drill.

The main functions and features of the jet during the breaking process are as follows. First, the jet cools the drill bit and prevents its destruction due to temperature rise, this extends the drill life. Second, during the advance and breaking stage the broken area and the free surface around the bit are formed in a way that reduces the effect of high confining pressures in the coal body. This speeds up the drilling. During the mixed and broken stage the jet enters the cracks and weakens the surface of the coal body, which expands fissures within the coal and improves the breaking efficiency of the bit. Third, at a certain angle between the nozzle and the drilling directions the jet can expand the drill hole and remove slag to reduce the slotting resistance to the tool. This also reduces the probability of a jamming event.

Relief of Pressure in Slotted Coal

Ground stress, gas pressure, and other factors create a circular stress concentration around the gas extraction borehole. Stress within the area of stress concentration is much higher than the original stress in the coal and rock. This leads to seam permeability. The single drill hole influences gas extraction results around it. Extraction is reduced and the penetration of gas through the coal body is blocked. This is shown in Fig. 2a. The slotted coal provides pressure relief and increases the permeability to take advantage of the high pressure jet slot and to agitate the coal and remove the coal powder. There are irregular slot like pressure relief spaces in the coal body. These can reduce stress concentrations in the borehole. Slotted coal around the stressed coal changes the situation as shown in Fig. 2b. Ground stress and gas gradient forces cause the coal body to move into the slot space, which creates a large number of tensile and shear cracks within the coal body. The pressure within the coal seam is thus relieved. The permeability of the coal seam is also increased. Macro slots and a large number of secondary cracks will, at the same time, provide a gas flow path that creates conditions for gas release. The slot relief lengthens the drainage gas range and will shorten the time for gas drainage. The pressure relief and increased permeability caused by concurrent drilling and slotting of the coal improve the results from a single hole gas extraction technique.

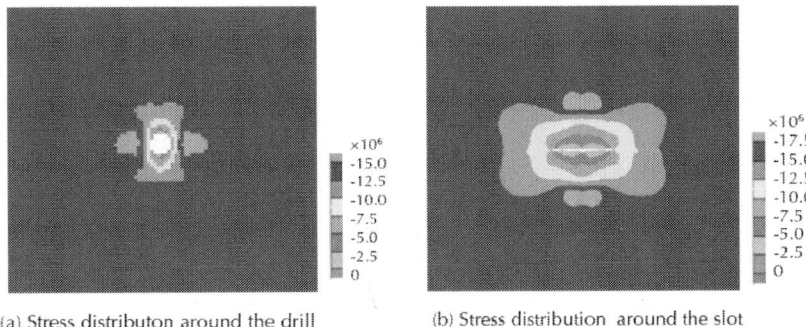

(a) Stress distributon around the drill   (b) Stress distribution around the slot

Figure 2: Stress changes before and after slotting.

The authors, and others, have done additional, in-depth studies on coal slotting relief and antireflection technology [1], [6] and [16]. This discussion highlights the double power concurrent breaking of the coal and explores the particulars of this method.

Influence of Confining Pressure on Coal Physical Properties

The micro-structural features of coal cause the physical properties of coal to change under the influence of high confining pressures. The coal breaking effect of a conventional drilling tool is affected by this and drilling becomes more difficult. Triaxial compression tests on different specimens from the same coal seam, shown in Fig. 3, allow the study of how confining pressure influences coal physical properties. An analysis of the change in ultimate breaking strength due to confining pressure has been reported [17] and [18]. The influence of confining pressure on coal physical properties consists mainly of the following:

- The confining pressure changes the ultimate strength and elastic modulus of the coal. This affects the deformation characteristics of the coal body.

- Low confining pressures (less than 10 MPa) result in ultimate strengths and elastic moduli that increase with increasing pressure. As the confining pressure continues to increase this trend changes. As the deformation of the coal mass increases the volumetric strain rate first increases but then decreases.

- The deformation and breaking of coal and rock are different under high and low pressures. When the confining pressure exceeds the elastic limit of the coal body the stress–strain curve shows significantly nonlinear characteristics and plastic deformation occurs.

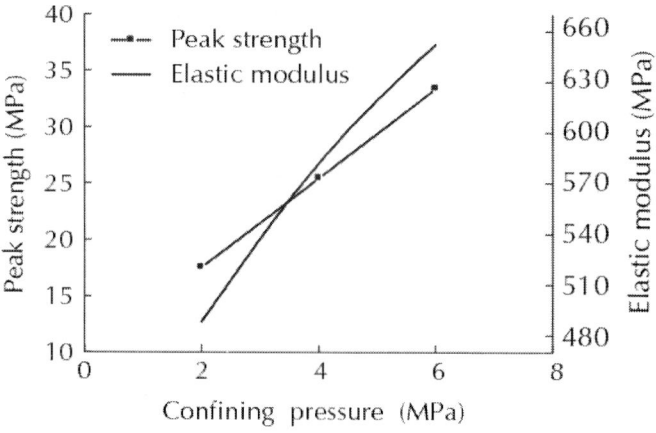

Figure 3: Ultimate strength and the elastic modulus of coal as a function of confining pressure.

This suggests that the impact of confining pressure on coal physical properties will mainly be seen in the peak strength and elastic modulus of the coal body. If the confining pressure is below the elastic limit, increases in confining pressure cause the ultimate breaking strength of the coal body to increase. Conversely, as the confining pressure drops the ultimate breaking strength of the coal body will decrease. Therefore, lowering the confining pressure will reduce the ultimate breaking strength of the coal body, thereby reducing the difficulty of breaking the coal with an alloy bit during drilling.

EXPERIMENTAL

Stress Evolution during Double Power Breaking of Coal

Analyzing changes in the stress during double power breaking was done by modeling with a Mohr–Coulomb numerical simulation using FLAC^{3D} code [19]. The size of the model was 4 m × 10 m × 4 m. A three dimensional "excavation" of the model was performed. The range of the simulated excavation was a shell 0.05 m wide. The bottom and boundary of the model were fixed. The stress in the vertical direction was 20 MPa and the stress in the horizontal directions was 10 and 5 MPa. Step by step excavation, based on the actual situation, was done in six steps. Each step was an excavation of 1 m and the results were noted at each one. The physical parameters and the grid for the model are shown in Table 1 and Fig. 4.

Table 1: Physical and mechanical parameters of the model coal and rock

Parameter	Bulk modulus (GPa)	Shear modulus (GPa)	Cohesion (MPa)	Tensile strength (MPa)	Friction angle (°)
Value	10	8	1	1	30

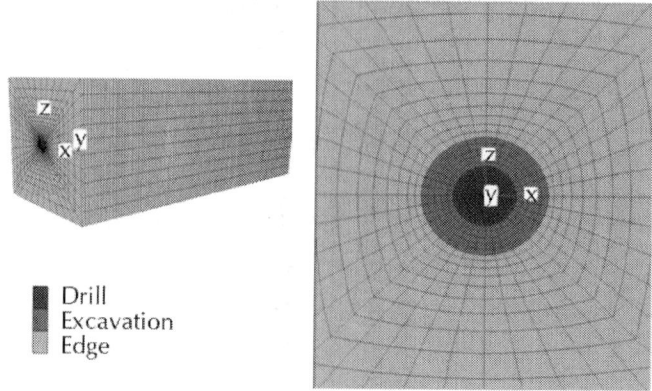

Drill
Excavation
Edge

Figure 4: Partial enlargement of the model grid.

The characteristic of dual coal drilling is that the advance and breaking effect of the high pressure jet forms an annular ring of a certain width and depth within the coal in front of the drilling bit. This can also cut off contact between original coal and the bit. This will create an "island" effect that reduces the wear on the bit during drilling and increases drilling capacity.

Fig. 5 and Fig. 6 show the stress concentration within the coal in front during drilling. The confining pressure influences the mechanical properties of the coal and rock. The stress concentration then results in the increase of the ultimate strength of the coal body, which in turn increases the difficulty of drilling through it. More difficult breaking reduces the drilling speed and can even lead to other technical problems. The advancing annular ring formed by the slotting jet cuts off the coal within the broken area from the original, surrounding coal. This creates a coal pillar that is fully pressure relieved and isolated. This coal will be easily damaged. Stresses along the drilling direction in the isolated coal pillar have a tendency to increase. The difficulty of breaking the coal pillar is influenced by changing the depth of the slot.

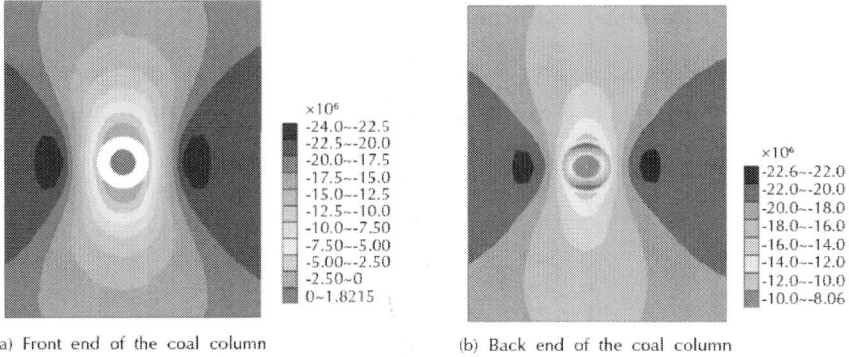

(a) Front end of the coal column   (b) Back end of the coal column

Figure 5: Stress distribution in planes perpendicular to the bore.

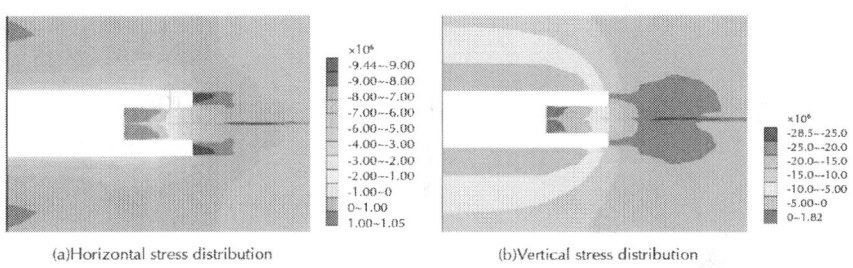

(a)Horizontal stress distribution   (b)Vertical stress distribution

Figure 6: Stress distribution along the bore.

Fig. 7 shows the stress at two different points on the central axis of the model during excavation. Fig. 8shows that during the formation of the annular ring the stress in the horizontal and vertical plane within the center of the residual coal pillar first drops slowly, then faster. Then the stress quickly increases and finally smoothly drops to show complete relief. The increasing stress segment shows that there would be a stress concentration within the coal pillar during drilling. But, along with the destruction from the jet, this stress concentration is quickly eliminated. Hence, the breaking strength of the coal in front of the drill can be effectively reduced only if the water jet creates an annular groove having a certain advance depth.

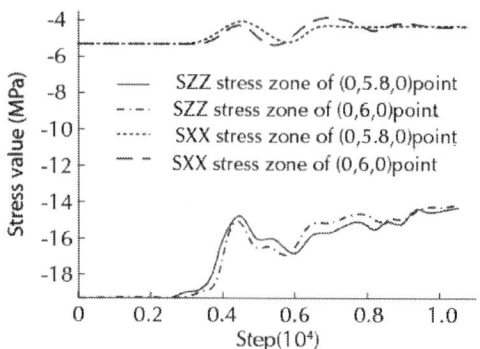

Figure 7: Stress in the center of the model.

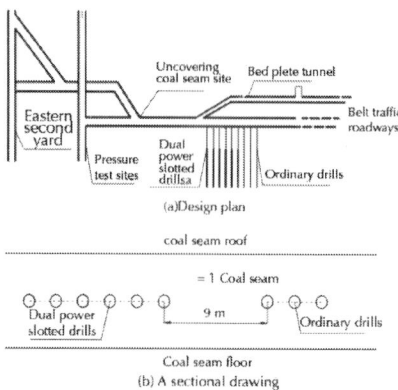

Figure 8: Layout of the experimental drillings.

Field Experiments

Experimental Construction and Design

The 11011 track roadway of the eastern second yard, Mengjin Coal Mine, was used as the site of a field application trial. The coal thickness of the 11011 track roadway is 4.6 m and the coal thickness of the number one drill field is 3.1 m. The gas pressure is 3.1 MPa 30 m distant from the 11011 track roadway. The real time gas level is not measured but the residual content was 13.96 m³/t after extraction for two months. The experimental drilling was designed as two groups of holes separated by 9 m with an in group interval of 3 m. The first group of six drillings was slotted with the dual power method. The other grouping of three drillings was an ordinary extraction done for comparison. Drilling along the coal seam was 1.5 m above the roadway floor. The drill diameter for both the dual power drill and the ordinary drill was 94 mm. The slotted depth in the dual power drill field was greater than 20 m and more than 20 m wide safety coal pillars were retained, as shown in Fig. 8. The design parameters of the slotted drillings are shown in Table 2.

Table 2: Parameters of the experimental drilling

Name	Diameter (mm)	Depth (m)	Slotted position (m)	Distance between holes (m)	Angle (°)
Parameter	94	>35	>20	3	3–6

Field Experiment

The experiment was conducted under secure conditions. The experimental observations of coal exhaust conditions, drill dynamic phenomena, and the changes in gas levels were recorded. The slotting process was an intermittent rotary cutting performed when the drill was retracted. There were six experimental drillings using dual power in the 11011 roadway. The experimental drill depth was increased significantly. During the experiment the gas concentration in the roadway was higher than for ordinary drilling. The softer coal at this site (f = 0.2), and the higher gas levels, cause phenomena such as coal burst, intermittent blocking, and hole spraying to occur often. The experimental drilling parameters and the conditions of the site are collected in Table 3.

Table 3: Parameters and site conditions during experimental drilling

Trial No.	Bit diameter (mm)	Drilling diameter (mm)	Drill depth (m)	Slot length (m)	Experimental condition	Maximum gas concentration (%)	Coal exhaust count (t)
1	94	128	43	20	Gas concentration becomes large after drilling 25 m, coal burst, and intermittent blocking appears when slotting	0.42	5–6
2	94	132	58	30	Serious coal burst during slotting, jet pressure is unstable. Intermittent blocking and drilling rig shake appear	0.58	7–8
3	94	127	55	25	Gas concentrations increased after drilling 30 m, holes block longer when slotting deeply	0.50	5–6
4	94	130	50	25	Serious coal burst when slotting at 40 to 50 m and hole spray frequently becomes severe. A single spray continues for a long time	0.69	7–8

5	94	131	56	32	Drilling rig shake appears when drilling at 40 m deep. Hole blocking is obvious, serious coal burst occurs when slotting	0.72	7–8
6	94	130	55	28	Gas concentration become large when slotting at 30 to 40 m with concomitant, serious hole spray	0.76	6–8
7	94	100	32		Bit binds when drilling 25 m deep	0.06	0.4
8	94	98	30		Hydraulic gauge of the drilling rig at a drill depth of 20 m, accompanied by noise	0.04	0.3
9	94	100	35		Drilling rig binds seriously at a drill depth of 24 m	0.05	0.4

The field experiments using the ordinary drill showed that it becomes difficult when drilling more than 20 m deep. Coal bursting and seriously drill binding occurred. Coal exhaust also became difficult for these trials. During dual power drilling the roadway gas concentration was observed to increase and the frequency of both hole spray and coal burst increased; These phenomena became more serious.

RESULTS AND DISCUSSION

An Investigation of Drill Depth and Coal Extraction

The field experiments were used to obtain the mean values of each parameter, ignoring extreme values. Drill depth, drilling diameter, and the coal extracted during dual power drilling and during ordinary drilling are shown in Table 4 and in Fig. 9.

Table 4: Drill depth, drilling diameter, and coal extraction observations

	Drill depth (m)	Drilling diameter (mm)	Coal extracted (t)
Dual power drilling	52.8	129.7	6.7
Ordinary drilling	32.3	99.3	0.4

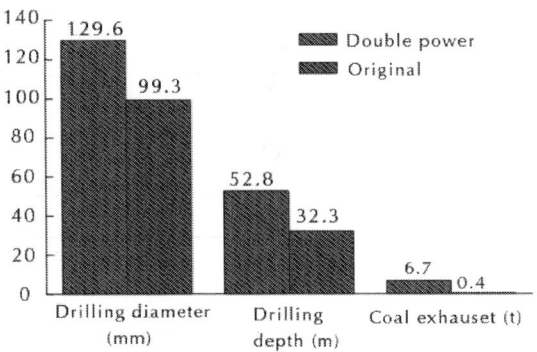

Figure 9: Drill depth, diameter, and coal extracted for the two methods.

Table 4 and Fig. 10 illustrate the different results for the two drill styles. The average drill depth during ordinary drilling was 32 m, while with dual power drilling it was 53 m. The drill depth increased by 72%. The drilling diameter increased 30%, from 99 to 130 mm, during dual power drilling, which enhances the ability of coal and gas to move away from the head of the drill and from the bore. Moreover, the slotting process of dual power drilling obviously increased the coal extraction, which was 18 times that seen during ordinary drilling. The pressure relief range and extraction range is expanded.

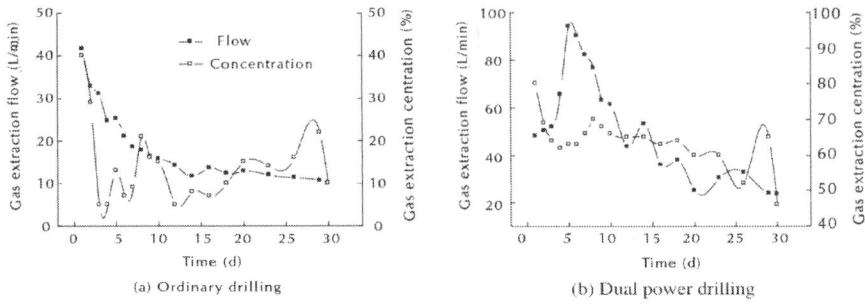

Figure 10: Gas extraction for one hundred meters of borehole.

Analysis of the experimental results shows that the coal seam at the experimental site had a high gas content, high ground stress, and soft coal. When ordinary drilling to some certain depth, the coal gas emission and confining pressure increased significantly. The difficulty with broken coal then increased. Furthermore, coal extraction from the drill head was difficult due to the increasing path distance. Dynamic phenomena during drilling could, and did, bind the bit, even in a lose drill pipe. But drilling with the dual power method relieved the high gas and confining pressure because of the breaking effect of the jet. This reduced difficulties during drilling and coal-body destruction. The drilling diameter was increased by the jet, the smaller bit and bigger hole favored coal extraction from the head and the drilling depth was subsequently improved. On the other hand, the damaging effects from the jet caused gas release to increase with a resulting higher gas concentration in the roadway.

An Analysis of Gas Extraction

When the series of experimental drillings were completed variation in the gas extraction volumes and rates were examined. The pure gas flow was measured accurately by observing gas extraction flows and concentrations on specific timing schedules. The measurement error was reduced by using single measurements and grouped statistics. Each drilling, both from the ordinary and dual power groups, had a flow meter and the concentration could be used to calculate the pure gas extraction flow. Summing and taking the mean allows a comparison between the two groups. The drill depth and extraction volume per group give the pure gas extraction volume for one hundred meters of drilling. Fig. 10 shows the gas extraction flow. The gas concentration for each of the two groups is also plotted.

Fig. 10 shows that the dual power drilling method can increase the rate and flow of gas extraction substantially. The highest pure gas extraction is 42 L/min for the ordinary method while the dual power method gave a maximum of 94 L/min, which is 2.3 times larger. The gas concentration remains at 60%, with small fluctuations, for the dual power method. The ordinary method gave gas concentrations below 30% with large fluctuations. During dual power drilling the drop in extracted gas flow is slower than with ordinary drilling. The coal bed permeability is improved after the slot from dual power drilling has been cut. The retrograde jet slotting of dual power drilling releases the internal energy of the gas and coal stress. Consequently, pressure relief is increased and coal fissures expand. The coal bed permeability of the affected areas increases substantially. The capacity for gas extraction increases significantly. So the efficacy of single hole gas extraction is improved.

A main measure of the effect of coal gas for pumping is the ratio of extracted gas volume to the total coal gas reserve. The total extraction is the integral of the instantaneous flow over the total extraction time. The area of the extraction curve gives the total gas extraction per month and the efficiency of extraction.

Consider an average gas extraction volume per hole over 30 days. The dual power drilling had a value of 720 m^3, while the ordinary drilling produced 235 m^3. The extraction rate also increased by a factor of two. The same extraction effect from the dual drilling method requires a fore pumping time significantly shorter than that required for ordinary drilling. The effect of pressure relief is obviously greater with dual power drilling, and the gas extraction from a single hole is improved.

CONCLUSIONS

- Pressure relief and permeability increasing techniques by slotting and drilling with a dual power drill are proposed. The mechanism of pressure relief and permeability increase is studied. The principles of dual power drilling are also analyzed. An analysis of coal stress evolution after dual power drilling showed that "islanding" occurs in the coal ahead of the drill bit during dual power drilling and that the coal failure strength is lower with less tool destruction during dual power drilling.

- Field experiments showed that compared to ordinary drilling the drilling depth during dual power drilling increased by 72%, the hole diameter expanded by 30%, and the amount of extracted coal was 17 times greater. Over 30 days the extracted gas flow from a single hole increased by 1.3 times in the dual power case. The gas concentration increased from less than 30%, with an unstable level, to more than 60%, with a stable level. The gas extraction efficiency increased by a factor of two times.

- Retrograde jet slotting during dual power drilling increased the coal extracted from the drill head. As a result, the internal gas and the coal stresses are released, and the pressure relief space is increased. Because the coal fissures expand the coal permeability in the affected areas increases substantially and the capacity for gas extraction increases significantly. The efficacy of single hole gas extraction is improved to the benefit of coal mine production safety.

ACKNOWLEDGMENTS

Thank for financial supports provided by the National Key Basic Research and Development Program of China (No. 2011CB201205), the National Natural Science Foundation of China (No. 51074161), and the Independent research of State Key Laboratory of Coal Resources and Mine Safety of China University of Mining & Technology (No. SKLCRSM08X03), and thank Mengjin Coal Mine, Yima Coal Mining Co. Ltd. for the help and support on field application and experiment of the technology.

REFERENCES

1.   Lin BQ, Yang W, Wu HJ. A numeric analysis of the effects different factors have on slotted drilling. J China Univ Min Technol 2010; 33(2):153–7.

2.   Lin BQ. Theory and technology of coal mine methane control.